P9-CAS-766

Affinity chromatography

a practical approach

Edited by

P D G Dean

Agricultural Genetics Co. Ltd., Cambridge Science Park,
Milton Road, Cambridge CB4 4BH, UK

W S Johnson

Biochemist, University of Liverpool, Liverpool, UK

F A Middle

Biochemical Consultant, P and S Biochemicals, Liverpool, UK

 **IRL PRESS**
Oxford · Washington DC

IRL Press Limited
P.O. Box 1,
Eynsham,
Oxford OX8 1JJ,
England

© 1985 IRL Press Limited

First published April 1985
First reprinting January 1986

*All rights reserved by the publisher. No part of this book may be
reproduced or transmitted in any form by any means, electronic or
mechanical, including photocopying, recording or any information storage
and retrieval system, without permission in writing from the publisher.*

British Library Cataloguing in Publication Data

Affinity chromatography : a practical approach.
 —(Practical approach series)
 1.   Affinity chromatography
 I.   Dean, P.D.G.   II. Johnson, W.S.
 III. Middle, F.A.   IV. Series
 543′.0892   QP515.9.A35

ISBN 0-904147-71-1

Printed in England by Information Printing, Oxford.

Preface

Discussions with leading people in the field of affinity chromatography revealed that there was a need for a methods book which would make available to the newcomer some of the favourite recipes used by the experts.

Affinity chromatography and immobilisation technology are inseparable. Both subjects are driven forward by their methods. While many spectacular advances have been made in the last 17 years, the period up to 1968 was almost as interesting in terms of the development of the applications of immobilised biochemicals.

The features of a practical methods book are important. (i) A precise account of the principal methods in the technique should be given. (ii) Novel developments should be included to add breadth and interest.

In this book we have tried to gather some of the laboratory procedures we know about, although this is limited to some extent to our needs over the years in the affinity chromatography group at Liverpool University. Many colleagues contributed to our research effort and bequeathed the valuable collection of practical notes on immobilisation techniques. This book is dedicated to their help and enthusiasm.

Since we began to collect these recipes, gene tailing was discovered by Steven Brewer and Helmut Sassenfeld at G.D.Searle. Their paper represents a milestone in affinity methods whereby the downstream selection method to isolate a gene product is organised at the level of DNA and not after the product has been expressed. Gene tailing uses N terminus polyarginine tails to render the protein exceptionally positively charged. In turn this means that cheap cellulose ion-exchange media can be used to separate the product from cell protein. Carboxypeptidase B is then used to remove the polyarginine tail and reconstitute the original protein. As a result, if a protein can be cloned, it should be possible to eliminate the need to develop sophisticated or costly purification methods.

Despite this key discovery, we believe that (i) affinity chromatography will remain a vital laboratory method and, furthermore, (ii) the basic concepts will continue to be taught in biochemistry courses.

We are indebted to Howard Long, Jeremy Walker, David Birch and Stephen Holmes for the library work and also for many of the diagrams, and to Sue Griffiths for secretarial and moral support over many years.

P.D.G.Dean, F.A.Middle and W.S.Johnson

Contributors

D.M.Abercrombie
Laboratory of Chemical Biology, National Institute of Arthritis, Diabetes and Digestive and Kidney Diseases, National Institutes of Health, Bethesda, MD 20205, USA

S.Allenmark
Department of Chemistry, Linköping University, S-581 83 Linköping, Sweden

S.Avrameas
Unité d'Immunocytochimie, Institut Pasteur, 28, rue du Dr. Roux, 75724 Paris Cedex 15, France

R.C.Barabino
Owens/Illinois, Inc., Biotechnology/Toxicology, One Seagate, Toledo, OH 43666, USA

E.Ber
Institute of Biochemistry and Biophysics, Polish Academy of Sciences, ul Rakowiecka 36, 02-532 Warsaw, Poland

B.Bomgren
Department of Chemistry, Linköping University, S-581 83 Linköping, Sweden

J.C.Bonnafous
Laboratoire de Biochimie des Membranes, ER CNRS 228, ENSCM, 8 rue de l'Ecole Normale, 34075 Montpellier, France

E.Boschetti
Pointet-Girard, 35 Avenue Jean-Jaures, 92390 Villeneuve-La-Garenne, France

I.M.Chaiken
Laboratory of Chemical Biology, National Institute of Arthritis, Diabetes and Digestive and Kidney Diseases, National Institutes of Health, Bethesda, MD 20205, USA

G.Dobrowolska
Institute of Biochemistry and Biophysics, Polish Academy of Sciences, ul Rakowiecka 36, 02-532 Warsaw, Poland

J.Dornand
Laboratoire de Biochimie des Membranes, ER CNRS 228, ENSCM, 8 rue de l'Ecole Normale, 34075 Montpellier, France

J.Favero
Laboratoire de Biochimie des Membranes, ER CNRS 228, ENSCM, 8 rue de l'Ecole Normale, 34075 Montpellier, France

E.A.Fischer
F.Hoffmann-La Roche & Co. Ltd., CH-4002 Basel, Switzerland

T.C.J.Gribnau
Technology Department, Organon International BV, P.O. Box 20, 5340 BH Oss, The Netherlands

J.K.Inman
Laboratory of Immunology, National Institute of Allergy and Infectious Diseases, National Institutes of Health, Bethesda, MD 20205, USA

M.H.Keyes
Owens/Illinois, Inc., Biotechnology/Toxicology, One Seagate, Toledo, OH 43666, USA

G.Krisam
Gambro Dialysatoren GmbH & Co. KG, 7450 Hechingen, Postfach 1323, FRG

C.Longstaff
Department of Pharmacology, Harvard Medical School, 250 Longwood Avenue, Boston, MA 02115, USA

J.-C.Mani
Laboratoire de Biochimie des Membranes, ER CNRS 228, ENSCM, 8 rue de l'Ecole Normale, 34075 Montpellier, France

I.Matsumoto
Department of Chemistry, Faculty of Science, Ochanomizu University, Otsuka, Bunkyo-ku, Tokyo 112, Japan

G.Muszyńska
Institute of Biochemistry and Biophysics, Polish Academy of Sciences, ul Rakowiecka 36, 02-532 Warsaw, Poland

E.Prusak
Biochemical Laboratory, Institute of Immunology and Experimental Therapy, Polish Academy of Sciences, Wroclaw, Poland

A.Szewczuk
Biochemical Laboratory, Institute of Immunology and Experimental Therapy, Polish Academy of Sciences, Wroclaw, Poland

R.R.Walters
Department of Chemistry, Iowa State University of Science and Technology, Ames, IA 50011, USA

D.J.Winzor
Department of Biochemistry, University of Queensland, St. Lucia, Brisbane, Queensland 4067, Australia

Contents

6. QUANTITATIVE CHARACTERISATION OF INTERACTIONS BY AFFINITY CHROMATOGRAPHY

D.J.Winzor

Abbreviations

AECM	aminoethylcarbamylmethyl
aPBA	*m*-aminophenylboronic acid
BNP	bovine neurophysin
BSA	bovine serum albumin
CBT	N,N'-carbonyl di-1,2,3-benzotriazole
CDI	N,N'-carbonyldiimidazole
CDT	N,N'-carbonyl di-1,2,4-triazole
CM	carboxymethyl
Con A	concanavalin A
CPG	controlled pore glass
DNP	dinitrophenyl
DTT	dithiothreitol
EDAC	1-ethyl-3-(3-dimethylaminopropyl)carbodiimide hydrochloride
EDTA	ethylenediamine tetraacetic acid
EEDQ	N-ethoxycarbonyl-2-ethoxy-1,2-dihydroquinoline
e.s.r.	electron spin resonance
FACS	fluorescence-activated cell sorter
FCP	2,4,6-trifluoro-5-chloropyrimidine
GGT	γ-glutamyltranspeptidase
HCG	human chorionic gonadotropin
Hepes	N-2-hydroxypiperazine-N'-2-ethanesulphonic acid
HMDA	hexamethylenediamine
n.m.r.	nuclear magnetic resonance
PBS	phosphate-buffered saline
PVA	polyvinylalcohol
SAMSA	S-acetylmercaptosuccinic anhydride
SPDP	N-succinimidyl-3-(2-pyridyldithio)propionate
TEMED	N,N,N',N'-tetramethylethylenediamine
TNBS	2,4,6-trinitrobenzenesulphonic acid
TSGT	thermal sol-gel transition
WGA	wheat germ agglutinin

HEALTH WARNING
USE OF CHEMICAL REAGENTS

Many of the chemicals used in the preparation of affinity chromatography media are toxic. Some are corrosive and others flammable. Wherever possible in this book specifically known hazards are highlighted. In general it is important when handling organic chemicals to abide by good safety standards, i.e., use of a laboratory coat, safety spectacles and disposable gloves. Particular care should be taken when carrying out procedures such as distillation, use of vacuum lines or bottled gases. The use of a fume hood is highly recommended.

CHAPTER 1

Matrix Preparations and Applications

1. INTRODUCTION

The ideal matrix for an affinity chromatographic support should have many properties. The following characteristics seem to be considered important by most authors: the matrix should be open and have a loose porous network; beads should be uniform in porosity and size, spherical and rigid; the beads should also be chemically and biologically inert, but at the same time derivatives must be easy to form preferably at room temperatures and in aqueous media; ideally the latter chemical derivatives should (i) be suited to ligand immobilisation (ii) be stable for some time and (iii) not disrupt either the matrix or the ligand, particularly if this is a protein; a modern matrix should also be able to withstand moderate pressures (5 bar); beads should be storable at low temperatures (e.g., $-30°C$). Clearly no one matrix available today fulfils all these criteria.

2. AGAROSE

The most popular matrix used for affinity chromatography is agarose. This is a purified linear galactose-containing (*Figure 1*) aerogel-xerogel hybrid colloid either isolated from agar or recovered directly from agar-bearing marine algae. Agarose has considerable gel strength and relative biological inertness and is therefore suitable for supports not only for affinity chromatography but also gel filtration. Agar and agarose are not synonymous and different agarose preparations may also vary widely with respect to their physico-chemical properties. Some agar-bearing seaweeds are shown in *Table 1*.

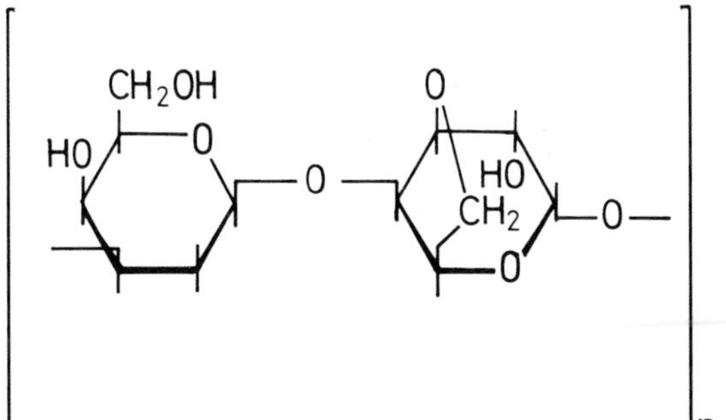

Figure 1. The repeating structural unit of agarose. The hydrogen atoms have been omitted for the sake of clarity. Reproduced from (1).

1

Table 1. The Biological Origins and Structural Features of Red Seaweed Galactans.

Galactan	Fraction	Sugar units
Agar Agar	Agarose	D-galactose 3,6-anhydro-L-galactose
	Agaropectin	
Carrageenan	Lambda	D-galactose-2-sulphate D-galactose-2,6-disulphate
	Kappa	D-galactose-4-sulphate 3,6-anhydro-D-galactose
	Iota	D-galactose-4-sulphate 3,5-anhydro-D-galactose-2-sulphate
Furcellaran		D-galactose D-galactose-4-sulphate 3,6-anhydro-D-galactose

2.1 Structure of Agarose

Agar consists of two fractions: a highly charged agaropectin fraction and a neutral agarose fraction. The 'neutral' agarose fraction is a repeating agarobiose unit consisting of alternating 1,3-linked β-D-galactopyranose and 2,4-linked 3,6-anhydro-α-L-galacto-pyranose moieties (*Figure 1*). It has since been shown that this concept of agarose is an over-simplification and that there does, in fact, exist a spectrum of species ranging from a very highly charged species through to a neutral agarose molecule. Although not always present concurrently, sulphate ester, methoxyl, ketal pyruvate and carboxyl groups can all appear on the agarobiose backbone.

Low organic sulphate content is frequently used as an indicator of agarose purity. This index of quality is based on the fact that many undesirable properties of agar are attributed to the ionic groups, of which sulphate is probably the major component. Commercial agarose beads contain up to 0.37% (w/v) sulphur and have a wide variation in the number of sulphate groups in this matrix as shown in *Table 2*. Removal of ionic groups by way of reduction with borohydride is a useful pre-treatment of the agar whenever any such groups interfere with a separation. In order to reduce agarose beads, they are treated with $NaBH_4$ in alkaline solution. For example, agarose (7 ml) is suspended in 1.0 M NaOH (5 ml) containing 2 mg/ml of $NaBH_4$. The reduction is continued for 2 h and the beads then washed with water.

The secondary structure of agarose, which consists of linear polysaccharide chains with no covalent cross-linking, leads to other problems including; (i) lack of thermal stability; (ii) shrinkage or (iii) swelling due to changes in ionic strength or dielectric constant of the medium; (iv) difficulty in freezing or drying and (v) ready solubility in the presence of denaturing or chaotropic ions. It is also impossible to use agarose with many organic solvents because the gel structure changes drastically and irreversibly under such conditions.

Table 2. Effect of Agar Type and Sulphur Content on Adsorption Capacity for Cytochrome c. Modified from ref. 2 with permission.

Gel	Adsorptive capacity	Sulphur (%)
Commercial agarose 6% (w/v)	0.08	0.11
ECD-agarose	0.01	0.02
Commerical agar 6% (w/v) (beads)	0.24	0.37
ECD-agar 6% (w/v) (beads)	0.06	0.05
Reduced ECD-agar 6% (w/v) (beads)	0.01	0.01

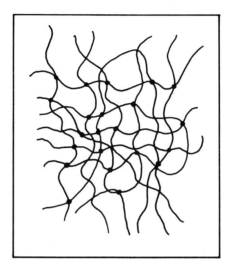

Figure 2. The structure of agarose.

The structural integrity, hardness, and porosity of the agarose gel depends on the secondary structure caused by non-covalent bonds (hydrogen bonds) between various agarose chains.

Figure 2 shows the proposed structure of the agarose network. The pores in agarose are large enough to be readily penetrated by proteins with molecular weights of several millions and yet agarose is strong enough to permit shaping into spherical particles with good flow characteristics. However, the stability of the pores is believed to depend on hydrogen bond formation between the three strands of the triple helix of agarose chains. Anything that is capable of disrupting these bonds will disrupt the entire network, and a solution of soluble agarose will result. The thermal instability of agarose is caused by the disruption of these bonds. Since hydrogen bonds are also disrupted by urea, guanidinium hydrochloride, chaotropic ions and certain detergents, any one of which may be desired as an eluent in affinity chromatography, it is necessary to cross-link agarose (with epichlorohydrin) which increases the stability of the matrix considerably, making it mechanically stronger and less affected by chaotropic reagents.

3

2.2 **Cross-linking of Agarose**

(i)   Suspend agarose beads (100 ml) in 1.0 M NaOH (1 litre) containing epichlorohydrin (10 ml) and NaBH$_4$ (0.5 g).

(ii) Heat the mixture to 60°C for 2 h with stirring. This can be conveniently done by attaching the flask containing the mixture to a rotary flask evaporator and immersing the rotating flask in a 60°C water bath. (**Caution:** do not apply a vacuum but secure the flask or tube to the drive shaft with a suitable clip). This method of stirring is preferred to use of a magnetic stirrer since the latter treatment may damage the beads, particularly if no rim is present on the stirrer bar.

(iii) After the reaction is complete, wash the beads with hot water until the washings are neutral. The beads may then be used without further additions but some carboxylate groups, created during the cross-linking, may be present. To remove these, autoclave the beads for 1 h at 120°C in 2.0 M NaOH (50 ml) containing NaBH$_4$ (0.25 g) then wash with 1.0 M NaOH (150 ml) containing 0.5% (w/v) NaBH$_4$.

(iv) Wash the gel with cold water, suspend in cold water and bring the suspension to pH 4.0 with acetic acid.

(v)   Store the beads at 4°C (do not freeze). If storage for more than a few weeks is required, prevent microbial growth by adding NaN$_3$ (final concentration 0.02% w/v).

2.3 **The Preparation of Agarose Beads**

Agarose for affinity separations is most commonly used in the form of beads, which are available commercially from several sources (*Table 3*). The preparation

Table 3. Some Commercial Sources of Supports Suitable for Affinity Chromatography.

Some Manufacturers and/or Suppliers of Affinity Adsorbents. The reader is referred to the relevant literature supplied with products in addition to the recipes found here since there are always refinements which are continually being made to these instructions. The addresses of these suppliers are to be found in Appendix I, p.30.

Amicon Corporation

Nylon-based affinity matrices (Matrex series 101 – 102), carboxylated nylons, cross-linked agarose beads (Matrex Gel series), immobilised dyes (Matrex Gel Dye Ligand Screening Kit) and phenylboronic acid, custom synthesised immobilised dye columns. Columns and ultrafiltration systems.

Bio-Rad Laboratories

Building blocks for affinity adsorbents including spacer molecules terminating in amino, carboxyl, thiol-phenylmercurial and activated carboxyl; agarose gels (Biol-Gel A series), CM-Biogel A, polyacrylamide gels (Bio-Gel P series). Chromatography equipment and columns.

Pierce Chemical Company

Building blocks for assembly of affinity supports based on Controlled Pore Glass (CPG). The following ligands are available: dextran, long chain alkylamines, phenylhydrazine, N-hydroxysuccinimide ester, *p*-nitrophenyl ester, carboxyl, thiol, stable diazonium salt, 8-hydroxyquinoline, lipoamide, CDI-activated supports: agarose, immobilised phenylboronic acid (Glycogel B), polystyrene beads derivatised with hydrazine, alkylamine and a wide variety of ligands.

Electro-nucleonics Inc.

Controlled Pore Glass (CPG) with various ligands available including; albumin, hexylamine, ethylamine and aminodipropylamine, adipic acid hydrazide and aminocaproate.

L'Industrie Biologique Française

Underivatised gels, agarose (Ultrogel series A), polyacrylamide-agarose copolymers (Ultrogel series AcA), magnetic varieties of the latter, glutaraldehyde activated gels, spacer arm gels, immobilised lectins, heparin, trisacryl gel series.

Koch-Light-Genzyme

Spheron series, polyacrylamide gels (Enzacryl series) including immobilised hydrazides, polythiolactones, polyacetals.

LKB-Producter AB

LKB market the IBF range of products but, in addition, chromatographic equipment, columns and a protein-compatible h.p.l.c. system.

E. Merck GmbH

Cellulose-based supports containing aminohexyl, succinyl aminododecyl, carboxymethyl hydrazide, immobilised phenylboronate, benzamidine, trypsin and trypsin inhibitor.

Miles Laboratories Inc.

Kits of hydrophobic columns C3 to C12, immobilised: hydrophobic-agaroses, agarose, 6-aminocaproyl-D-tryptophan methyl ester, soybean trypsin inhibitor, Gly-Gly-L-Tyr(O-Bz)-L-Arg, L-lysine, L-thyroxine, L-3,5,3-triiodothyronine, 6-aminocaproylfucosamine, *p*-aminophenyl mercuric acetate, 5'-(4-aminophenyl-phosphoryl-uridine-2'(3')-phosphate), casein, L-tryptophan, L-tyrosine, L-phenylalanine, concanavalin A, fucose binding protein, wheat germ lectin, soybean agglutinin, *Lens culinaris* haemagglutinin A, *Lens culinaris* haemagglutinin B, *Ricinus communis* agglutinin, immunochemicals, poly(L-lysine), succinyl-poly(L-lysine), succinyl-poly(DL-alanine)-poly(L-lysine), adipic hydrazide, polyacrylic hydrazide, succinyl polyacrylic hydrazide, bromoacetyl-cellulose, enzymes.

Marine Colloids Inc.

Agarose beads (SeaSep AC beads).

Pharmacia − P.L. Biochemicals Inc.

2%, 4%, 6%, 8%, 10%, (w/v) agaroses; immobilised: 1,6-diaminohexane, concanavalin A, cysteamine, haemoglobin, hexane, adenosine, mono-, di- and tri-nucleotides, coenzyme A thiol, thiol coupler, hexanoic acid, N-hydroxysuccinimide ester, soya bean and lima bean trypsin inhibitors, poly(A,C,G,I,I-C,U)-cellulose, cellulose-immobilised nucleotides, DNA (denatured and native), oligo(dT); Sepharose series; Superose series; Sepharose CL series; Sepharose ion-exchangers; Sephadex series; Sephadex LH-20; Sephacryl series; Sephadex ion-exchange series; Blue Dextran, Red and Blue Sepharose CL-6B; building blocks for affinity systems including: aminohexyl; carboxyhexyl; CNBr-activated Sepharose 4B; CNBr-activated Sepharose 6MB; epoxy-activated Sepharose 6B; hydrophobic ligands: octyl-Sepharose 4B; immobilised lectins; phenyl-Sepharose CL-4B; protein A-Sepharose CL-4B; chromatographic columns, pumps and equipment; fast protein liquid chromatography system.

Rohm Pharma GmbH

Eupergit C, epoxy-activated supports.

Serva Feinbiochemica GmbH & Co.

Agarose bound: 1,2-diaminothane, 3,3'-diaminodipropylamine, succinylated analogues, aminopropyl-*p*-aminobenzyl-cellulose, *p*-aminobenzoyl-3,3'-diaminodipropylamine-agarose (Servachrom series).

Sigma Chemical Company

Agarose immobilised: nucleotides, adipic dihydrazide, ε-aminocaproic acid, N-(*p*-aminophenyl)-oxamic acid, coenzyme A, concanavalin A, cysteamine, diaminohexane, fetuin, haemoglobin, α-lactalbumin, polynucleotides.

This list is not comprehensive: full details of the product ranges should be obtained from each company.

Table 4. Experimental Data for the Preparation of Agarose Gel Beads by Suspension Gelation.

The volume of water in which the agarose is dissolved in all experiments is 100 ml. The right column gives the size of the majority of the agarose spheres. The stabilisers may be obtained from Atlas Chemical Industries Inc., Chemicals Division, Wilmington 99, Delaware, USA.

Agarose conc. (g agarose added to 100 ml water)	Toluene (ml)	CCl₄ (ml)	Stabiliser	Amount (g)	Stirring speed (r.p.m.)	Sphere size mesh, US standard
6	450	150	Sorbitan sesquioleate	4.5	1700	100 − 170
8	445	155	Polyoxyethylene sorbitan monostearate	5	1500	60 − 100
10	440	160	Polyoxyethylene sorbitan monostearate	15	1700	170 − 300
12	430	170	Polyoxyethylene sorbitan monostearate	25	1700	170 − 300

of agarose beads with a range of sizes has been described by Hjerten (3).

The process described by Hjerten is complex. However, the following factors are important in determining bead quality: the nature and amount of the stabiliser, the oil phase, stirrer speed, concentration of the agarose, volumes and densities of water and oil phases, shapes and relative dimensions of the stirrer and the container all affect the size and shape of the gel beads obtained (see *Table 4*).

2.4 **Method for Preparation of Agarose Beads**

(i)   Equip a 1 litre round bottomed-flask with a ground glass joint with a stirrer blade made of anode-oxidised aluminium. During gelation, the viscosity of the suspension changes continuously. It is important, therefore, to ensure a relatively constant stirring speed by using a variable speed motor.

(ii)   Dissolve the agarose by boiling in the round-bottom flask. For agarose concentrations above 3% (w/v) it is necessary to autoclave the agarose at 2 bar (30 p.s.i.) in order to dissolve the agarose.

(iii)   Evacuate the solution before heating. Several attempts may be needed before total solution is achieved, particularly with agarose concentrations greater than 10% (w/v).

(iv)   Pre-heat the organic phase (see *Table 4*) containing the stabiliser to 50°C in a water bath.

(v)   Add the organic phase to the agarose and start the stirrer.

(vi)   After 1 min, cool the mixture by surrounding the flask with cold water.

(vii)   After 5 min wash the beads on a sintered glass funnel. Wash the beads in ether (3 x 200 ml) in order to remove the organic solvent.

(viii) Add water (1 litre) and decant off the ether layer.

(ix)   Remove the ether dissolved in the aqueous phase by placing the contents of the flask in a Buchner flask and evacuating the latter with a water pump.

(x)   Further wash the beads on a funnel and sieve to remove fines.

(xi)   Examine the beads under the microscope after dyeing with Cibacron Blue (as described in Chapter 6) for uniformity of shape and distribution of bead diameter.

2.5 Gel Properties of Superose 6B, an Agarose Gel for High Performance Affinity Separations

Superose 6B is an agarose matrix which is highly cross-linked, resulting in a very rigid gel. Both the cross-linking and narrow particle size ($20-40$ μm) contribute to its performance as a chromatography support for high performance separations. Typically according to the manufacturers, (Pharmacia), a column (1.6 x 60 cm) operated at a flow rate of 0.3 ml/min produces a back pressure of less than 1 bar (15 p.s.i.). The fractionation range (molecular weight) is $10^4 - 4$ x 10^6 for globular proteins and $10^4 - 10^6$ for polysaccharides.

Superose 6B shows a very low degree of non-specific adsorption. Protein recoveries are usually greater than 80%. Superose 6B can be used in aqueous media in the pH range $3-14$. The gel is stable in solutions of chaotropic reagents (up to 6.0 M). Solutions of ethylene glycol, ethanol, dimethylformamide, tetrahydrofuran, acetone, chloroform, dichloromethane and dichloroethane-pyridine (50:50) may all be used, but not solutions of urea above 6.0 M.

Superose 6B can be repeatedly autoclaved at pH 7.0, 120°C without significant changes in porosity and rigidity. The gel can be used up to a pressure of 4 bar (60 p.s.i.). Anionic, cationic and non-ionic detergent solutions may be used without harming the gel.

Before packing columns of Superose, the user is advised to filter and degas all solutions. Samples should also be treated in the same fashion. Pharmacia supply useful further information relating to this gel.

2.6 Magnetic Agarose (Magnogel)

Magnogel A4R is a support composed of agarose (4%, w/v) cross-linked with epichlorohydrin. As with Magnogel AcA 44 (see below), its magnetic nature results from the incorporation of 7% (w/v) Fe_3O_4 in the interior of the gel beads. In addition to being magnetic, Magnogel A4R is heat resistant and stable in the presence of dissociating agents as a result of cross-linking. It has applications in situations where column operation is not favoured, e.g., in viscous solutions or in the presence of partly soluble materials such as cell debris.

3. CELLULOSE

Cellulose consists of linear polymers of β-1,4-linked D-glucose units with occasional 1,6-bonds.

3.1 Beaded Celluloses: Introduction

A wide range of derivatised celluloses have been described and their applications have generally been widespread throughout protein purification methodology. However, several limitations of cellulose have always been recognised, not only in general purification methods such as ion-exchange chromatography but also in affinity chromatography. These problems are derived from the unsuitable physical structure (lack of porosity) and from the unsatisfactory geometrical shape of the individual particles. Furthermore, extensive microcrystalline areas

7

within the matrix exacerbate the problem. Recently most problems related to the use of the older fibrous, powdered celluloses have been solved by the introduction of novel forms, one of which is both porous and spherical.

3.2 Preparation of Beaded Cellulose

A number of procedures for the preparation of beaded cellulose have appeared. All are based on a common principle, and include the following steps:

(i)   formation of a liquid phase containing cellulose or a cellulose derivative;
(ii)   shaping the liquid to form spherical droplets;
(iii)   solidification of the liquid droplets;
(iv)   regeneration of cellulose into solid spherical particles;
(v)   final washing.

In order to distribute cellulose into individual liquid droplets, the fluid is either extruded through suitably sized apertures or dispersed in a medium which does not mix with the cellulose-containing liquid phase (4 – 17). The particle size is controlled by the manner in which the fluid passes through the nozzle, by the efficiency of mixing during dispersion, or by the addition of surface-active compounds.

An important step in the preparation is the solidification of liquid particles, i.e., completion of the sol-gel transition. This is carried out in such a way as to avoid deformation of the spherical shape and adhesion of the individual particles to yield agglomerates. The procedures used to achieve the sol-gel transition involve salt and acid regeneration baths as used in the production of cellulose fibres; also, the composition of the dispersed phase is changed so as to reduce the solubility of the cellulose component, the temperature is decreased in order to achieve solidification of the melt (in the case of cellulose acetate) or to reduce the solubility, and chemical cross-linking is used, especially by using epichlorohydrin in an alkaline medium. The preparation of spherical celluloses is completed by various additional procedures which result in a more porous structure. These procedures are designed to complete the regeneration of the cellulose and, in addition, remove decomposition products by washing.

3.3 Properties of Beaded Cellulose

Stamberg and co-workers have produced beaded cellulose using the thermal sol-gel transition (TSGT) process which yields uniform spherical beads. The geometry and particle size distribution of the beads have been described elsewhere (18). Beaded cellulose is a highly porous spherical product which is composed of pure regenerated cellulose (15%) and water.

With its porous structure, the swollen beaded cellulose resembles vinyl copolymers of the macroporous (macroreticular) type. In the coagulation and regeneration of cellulose from xanthate solutions, a microheterogeneous structure is formed consisting of regions of crystalline order interconnected by amorphous material. Owing to this microheterogeneity, spherical macroparticles of beaded cellulose are much more rigid than particles of homogeneous polysaccharide gels which arise, for instance, by the chemical cross-linking of the individual polymer chains with epichlorohydrin. The introduction of cross-linking groups or other substituents into macromolecules may reduce the formation of crystalline domains. For this reason, beaded cellulose is claimed to be more stable mechanically than, for example, dextran gels of comparable porosity.

The porous structure of beaded cellulose may be modified during production to yield tailor-made spherical cellulose carriers. Changes in the porous structure are used to carry out these modifications by direct removal of water. Drying gradually reduces the amount of water which cellulose can retain in its pores on subsequent re-swelling. This is due to the contraction of the skeleton and to the formation of new hydrogen bonds and other structure-forming bonds. Drying leads to the formation of larger, new microcrystalline domains.

The porous structure may be varied by the solvent replacement method followed by drying or re-swelling. Pores filled with water in the initial state can be filled with any solvent by exchange without shrinkage and a corresponding decrease in porosity can be engineered. According to the type of solvent from which cellulose is eventually dried, products of various true porosity (dry-state porosity) are formed with a characteristic macroreticular structure. For example, beaded cellulose dried from benzene has a porosity value of 83% (vol % of pores) and a surface area of 340 m²/g. Drying from solvents of higher polarity yielded products with lower porosity; no porosity was demonstrated in the dry state for the product dried from water.

After repeated water swelling of cellulose samples dried from solvents of various polarity, different water regain values can be obtained. By altering the conditions of drying, it is possible to prepare products with porosity values varying in the dry state from 0 to 83% and in the water-swollen state from 50 to 90%.

It is known that cellulose exceeds some other synthetic polymers in its chemical reactivity. Many reactions proceed with particular ease owing to the porosity of the skeleton and the accessibility of reaction sites. In some cases, it is advisable to modify the porosity by adjusting the content of the swelling liquid. This can produce an undesirable dilution of the ligand immobilisation reaction solution. To overcome this problem, it may be appropriate to fill the inter-particle volume with an inert medium, e.g. a hydrocarbon, thus directing the activation reaction into the intra-particle domain.

Beaded cellulose is theoretically a useful support matrix for affinity chromatography, because of its regular spherical shape, its high porosity and good mechanical stability and its pronounced hydrophilic character. These applications have yet to be demonstrated. In one report, dye ligand chromatography is claimed to be possible (19) whilst in a second (20), these observations could not be confirmed. On the other hand, a number of enzymes have been successfully immobilised onto beaded cellulose.

4. DEXTRAN

Dextran is an α-1,6-linked glucose polymer made by, and produced from, *Leuconostoc mesenteroides*. The commercial product, Sephadex, which is a gel filtration medium (see *Table 3*), is prepared by cross-linking dextrans with epichlorohydrin and is available in bead form. Hydroxyl groups on the polysaccharide backbone make this matrix more hydrophilic than agarose. Sephadex may be reversibly swollen from the dried state with no significant changes in the chromatographic properties of the gel after repeated drying and re-swelling cycles. Sephadex is chemically very stable, with the gel surviving intact after 2 months in 0.25 M NaOH at 60°C or 6 months in 0.02 M HCl (21). Furthermore, wet Sephadex can be autoclaved without altering its properties. However, prolonged exposure to oxidising agents can increase the carboxyl group content.

Cross-linked dextran derivatives ought to make an ideal matrix for affinity chromatography, but the low degree of porosity of the mechanically more stable forms (G10 to G75) outweighs any real advantages. A lower degree of cross-linking results in rather fragile gels. Activation of the dextran gels by most common methods can further cross-link the gel and makes the gels virtually impenetrable, even to enzymes of low molecular weight.

5. POLYACRYLAMIDE

Polyacrylamide gels are composed of a hydrocarbon skeleton on to which carboxyamide groups are bound. *Figure 3* shows a linear polyacrylamide chain.

Polyacrylamide gels are stable over the pH range 1 − 10 and they can be used with most common eluents. They do not contain charged groups, and so ion exchange interactions with chromatographed substances are minimal. They are biologically inert and, because of their inert structure, they are not attacked by microorganisms. As the gel particles adhere strongly to clean glass surfaces, Inman and Dintzis (22) recommend the use of siliconised glass or polyethylene laboratory vessels. The lack of mechanical stability of polyacrylamide is a disadvantage which limits its usefulness as a matrix for affinity chromatography.

On reaction with suitable compounds, polyacrylamide gels can be converted into solid carriers suitable for the binding of a series of ligands (22). Thus the aminoethyl derivatives can be prepared using a large excess of ethylenediamine at 90°C, and hydrazide derivatives by using an excess of hydrazine at 50°C. Aminoethyl derivatives of polyacrylamide gels can be converted into their *p*-aminobenzamidoethyl derivatives by reaction with *p*-nitrobenzoylazide in the presence of N,N'-dimethylformamide, triethylamine and sodium thiosulphate.

One of the main producers of polyacrylamide gels is Bio-Rad Laboratories, whose product is marketed under the trade-name Bio-Gel P, prepared by co-polymerisation of acrylamide and N,N'-methylenebisacrylamide. Bio-Gel P is produced with various pore sizes which range from Bio-Gel P-2 with a molecular weight exclusion limit of 1800 up to Bio-Gel P-300 with a molecular weight exclusion limit of 400 000. All brands are available in 50 − 100, 100 − 200, 200 − 400 and 400 mesh size. In addition to these gels, Bio-Rad Labs produce ion-exchange

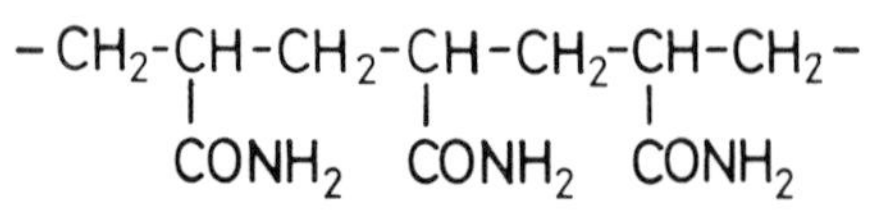

$$- CH_2-CH-CH_2-CH-CH_2-CH-CH_2 -$$
$$CONH_2 \quad CONH_2 \quad CONH_2$$

Figure 3. A linear polyacrylamide chain

derivatives of the gels, for example the weakly acidic cation exchanger Bio-Gel CM, and also intermediates for affinity chromatography, such as the aminoethyl and hydrazide derivatives of Bio-Gel P-2 and P-60.

A commercial variation on pre-activated polyacrylamide is the Enzacryl series (Koch-Light, Colnbrook, UK). These gels are designed for enzyme immobilisation. Enzacryl AH is a hydrazide derivative of polyacrylamide gels, and Enzacryl AA is a polyacrylamide gel containing aromatic acid residues. Enzacryl Polyacetal is a co-polymer of N-acryloylaminoacetaldehyde dimethyl acetal with N,N′-methylenebisacrylamide, which binds proteins through their NH_2 groups.

6. TRISACRYL

by E.Boschetti

Trisacryls are a range of gels manufactured by IBF Réactifs in Paris and available from LKB. They are synthetic gels which serve as gel filtration media, ion exchange and affinity chromatography supports. They are derived from the polymerisation of the unique monomer N-acryloyl-2-amino-2-hydroxymethyl-1,3-propane diol (*Figure 4*). The resulting polymer is illustrated in *Figure 5*.

The main characteristic of these macromolecules is that they bear three hydroxymethyl groups and one alkylamide group for each principal repeating unit. Because of these chemical functions, the polymers are very hydrophilic and suitable for the separation of biological macromolecules, especially proteins, and cells.

The amino-2-hydroxymethyl-2-propanediol residues create a micro-environment that favours the approach of hydrophilic solutes (proteins) to the polymer surface. The polyethylene backbone is buried underneath a layer of

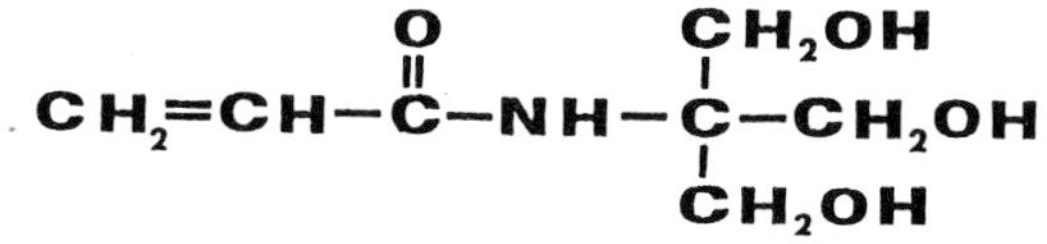

Figure 4. The monomer used to make trisacryl gels.

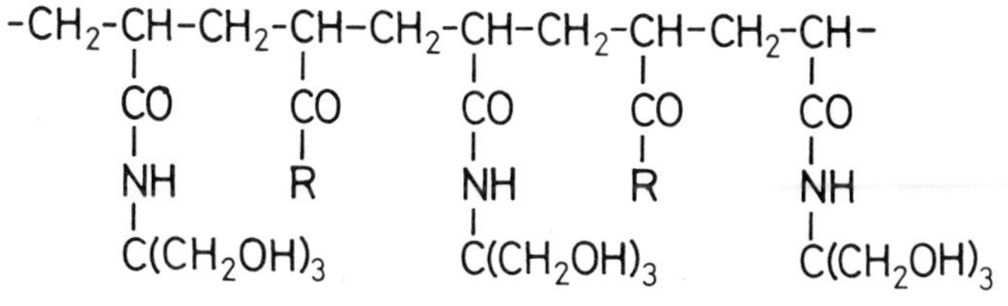

Figure 5. The structure of trisacryl ion exchangers.
CM-trisacryl, R = COOH
DEAE-trisacryl, R = NH-$(CH_2)_2$-NEt_2
SP-trisacryl, R = NH-CMe_2-CH_2-SO_3H
Unsubstituted trisacryl, R = NH_2

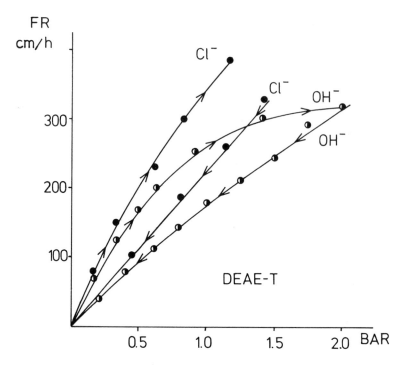

Figure 6. The response of flow-rate to increasing pressure for DEAE trisacryl gels.

hydroxymethyl groups. This type of matrix has an obvious advantage over polyacrylamide or hydroxymethyl methacrylate-based supports, which have a pronounced hydrophobic character. In addition, it is claimed that trisacryl beads do not interact with cells, which may be because they possess no saccharide structures, unlike classical chromatography matrices. Trisacryls are likely to find important uses in high performance systems since they are resistant to moderate pressures (up to 3 bar) see *Figure 6*.

Trisacryl gels are very hydrophilic, non-biodegradable, and stable to low (−20°C) and high (121°C) temperatures and denaturing agents. They are also stable to acidic pH but less stable to high pH because of the slow hydrolysis of the amide linkage. *Table 5* lists some of their more important properties.

The exclusion limits of ion exchangers as determined by the chromatographic behaviour of globular macromolecules are high. *Figure 7* shows a comparison between the behaviour of trisacryl matrix involved in the synthesis of ion exchangers and a gel filtration medium Ultrogel AcA 34. It can be seen that there is little effect of molecular weight on the elution characteristics of the trisacryl gel.

The capacities of trisacryl gel ion exchange media are dependent upon pH, ionic strength, temperature, the nature of the buffer and the sample concentrations. *Table 6* shows some of these properties.

Table 5. Some Properties of Trisacryl Gels.

	CM-trisacryl	*DEAE-trisacryl*	*SP-trisacryl*
Bead size	$40-80\ \mu m$	$40-80\ \mu m$	$40-80\ \mu m$
Exclusion limit (mol. wt.)	$>> 10^7$	$>> 10^7$	$> 10^7$
Ionised group density	$200\ \mu E/ml$	$300\ \mu E/ml$	$230\ \mu E/ml$
pK of carboxyl group	4.7	–	–
pK of amino groups	–	6.2, 10.7	–
pK of sulphonate	–	–	1.0
pH stability	$0-13$	$0-13$	$0-13$
Stability to detergents	excellent	excellent	excellent
and denaturing agents	excellent	good	excellent
Heat stability	excellent	excellent	excellent
Resistance to microbial degradation	excellent	excellent	excellent
Volume changes due to pH and ionic strength	negligible	negligible	negligible

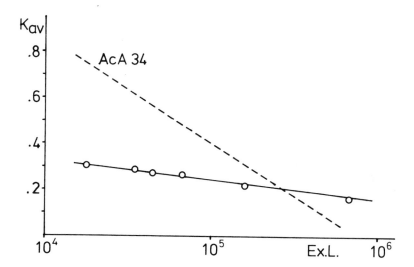

Figure 7. The variation of K available (K_{av}) with molecular weight for trisacryl ($\bigcirc$——$\bigcirc$) and Ultrogel AcA 34 (----).

Table 6. The Capacities of CM-Trisacryl and DEAE-Trisacryl.

	CM-trisacryl	*DEAE-trisacryl*
Total capacity	$200\ \mu E/ml$	$300\ \mu E/ml$
Capacity for bovine serum albumin[a]	–	105 mg/ml
Capacity for human haemoglobin[b]	95 mg/ml	85 mg/ml
Capacity for cytochrome c[c]	95 mg/ml	–

[a]Determined in 0.05 M Tris-HCl, pH 8.0, with 10 mg/ml solution of bovine albumin.
[b]Determined (for CM) in 0.05 M acetate buffer, pH 5.0; (for DEAE) in 0.05 M Tris-HCl buffer, pH 9.0. A solution of 5 mg/ml haemoglobin was used.
[c]Determined in 0.05 M acetate buffer, pH 5.4. A solution of 3 mg/ml of cytochrome c was used.

6.1 **Preparation of Trisacryl Polymer in Bead Form**

Trisacryl polymer may be prepared on a laboratory scale according to the method described in French patent no. 7,702,391.

(i)   Prepare a 1 litre volume of an aqueous solution of monomers by dissolving N-acryloyl-2-amino-2-hydroxymethyl-1,3-propanediol (330 mg/ml) with N,N'-diallyltartradiamide as a cross-linking agent (40 mg/ml).

(ii)   Adjust the temperature of the solution to 55°C in a water-bath, and when the mixture is equilibrated, add ammonium persulphate (120 mg) and N,N,N',N'-tetramethylethylenediamine (1.6 ml).

(iii)   Emulsify this mixture immediately in paraffin oil (2 litres) stirring thoroughly. Under these conditions, an exothermic polymerisation occurs to form spherical solid particles.

(iv)   Recover the particles by decantation, wash with a 1% (v/v) aqueous Triton X-100, sieve to obtain the fraction between 40 and 80 μm and finally wash with 1.0 M sodium chloride, water and store at 4°C in 0.02% (w/v) NaN$_3$ (final concentration).

6.2 **Introduction of Amino Groups on Trisacryl Beads (Trisacryl-NH$_2$)**

Treat trisacryl beads (30 ml) with 25% (w/v) zinc tetrafluoroborate (22.5 ml) and epichlorohydrin (60 ml), for 3 h at 80°C, with gentle agitation. After cooling the reaction mixture, rinse the beads with distilled water and resuspend in 2.0 M ammonia-ammonium chloride, pH 9.0 (100 ml). After 16 h gentle agitation at room temperature, wash the beads thoroughly with distilled water and store at 4°C in 0.02% (w/v) NaN$_3$.

7. HYDROXYALKYL METHACRYLATE GELS

Hydrophilic hydroxyalkyl methacrylate gels, introduced by Wichterle and Lim (23), may be prepared by polymerisation of a suspension of hydroxyalkyl esters of methacrylic acid and alkylene dimethacrylate by varying the ratio of the concentrations of monomer and inert components (24). The number of reactive groups, the porosity and the specific surface area of the gel may be changed within broad limits. The gel has the structure shown in *Figure 8*.

The gels are produced under the trade-name Spheron both by Lachema (Brno, Czechoslovakia) and Realco Chemical Co. (New Brunswick, NJ, USA). The gels form regular beads with excellent chemical and physical stabilities. They withstand chromatography under pressure well and do not change their structures after heating for 8 h in 1.0 M sodium glycolate solution at 150°C or after boiling in 20% (v/v) HCl for 24 h. They are biologically inert and, like acrylamide gels, are not attacked by microorganisms. They can be employed in organic solvents. The hydroxyl groups of the gel possess properties analogous to those of agarose.

8. POLYACRYLAMIDE-AGAROSE GELS

The AcA series of Ultrogels are polyacrylamide-agarose gels (see *Figure 9*) with different and standardised porosities according to the type. There are five types of AcA gels with different porosity all comprising a three-dimensional

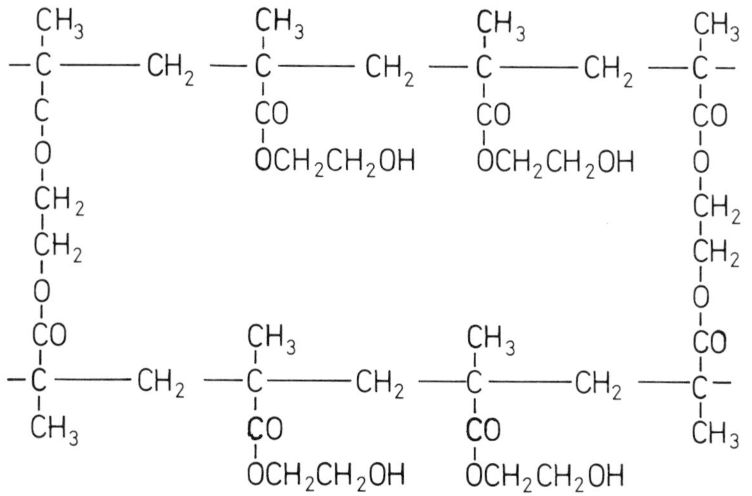

$$
\begin{array}{cccc}
CH_3 & CH_3 & CH_3 & CH_3 \\
-C\!-\!\!-\!CH_2\!-\!C\!-\!\!-\!CH_2\!-\!C\!-\!\!-\!CH_2\!-\!C\!- \\
C & CO & CO & CO \\
O & OCH_2CH_2OH & OCH_2CH_2OH & O \\
CH_2 & & & CH_2 \\
CH_2 & & & CH_2 \\
O & & & O \\
CO & CH_3 & CH_3 & CO \\
-C\!-\!\!-\!CH_2\!-\!C\!-\!\!-\!CH_2\!-\!C\!-\!\!-\!CH_2\!-\!C\!- \\
CH_3 & CO & CO & CH_3 \\
& OCH_2CH_2OH & OCH_2CH_2OH &
\end{array}
$$

Figure 8. Structure of hydroxyalkyl methacrylate gel.

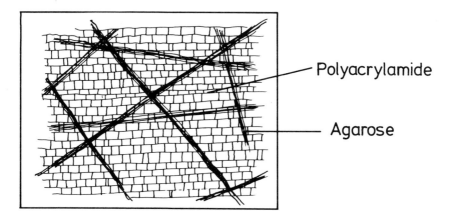

Figure 9. Schematic representation of Ultrogels AcA matrix.

polyacrylamide lattice enclosing an interstitial agarose gel. The fractionation range of these gels is between 1000 and 1 200 000 daltons. These gels have the unique characteristics of bearing two types of chemical groups which can be modified to immobilise ligands: the hydroxyl groups of agarose and the amide groups of polyacrylamide. Ultrogel beads are more rigid, and hence less compressible than conventional gel media and thus permit higher flow-rates.

Magnogel AcA 44 is a magnetised support, derived from Ultrogel AcA 44. The polyacrylamide-agarose beads contain 7% (w/v) iron oxide (Fe_3O_4) internally, thus enabling them to be attracted by a magnetic field. This gel is used for microassays and preparations of products in heterogeneous media containing particles in suspension. In the case of microassays, this technique avoids the numerous centrifugation steps which accompany conventional gel assays.

15

9. IMMOBILISATION OF ENZYMES IN POLY(VINYL ACETATE-CO-ETHYLENE) TUBES

9.1 Introduction

Manecke (25) has described various reactive carriers for the immobilisation of enzymes using as insoluble starting materials polymeric products which contain polyvinylalcohol: polyvinylalcohol gels cross-linked with terephthalaldehyde, hydrolysed beads of cross-linked poly(vinyl acetate), poly(vinyl acetate-co-ethylene) tubes coated with polyvinylalcohol and mainly polyvinylalcohol-containing synthetic wood pulp. Other supports used in affinity chromatography include nylon, polystyrene, paper, sodium alginate and insolubilised proteins.

Poly(vinyl acetate-co-ethylene) tubes are available from Boehringer Mannheim.

9.2 Methods for Preparation of Reactive Tubes made from Vinyl Acetate-Ethylene Co-polymers

All numbers (I – VIIIa) in this section refer to *Figure 10*.

9.2.1 *Activation of Tubes made from Vinyl Acetate-Ethylene Co-polymers*

Fill a 500 cm length of poly(vinyl acetate-co-ethylene) tube (0.2 cm bore, I in *Figure 10)* with 20% (w/v) NaOH (25 ml). Incubate the tube at 40°C for 2 h by recirculating the solution with a peristaltic pump.

9.2.2 *Derivatisation of the Inside Tube Surface with Epoxypropoxybenzaldehyde to III*

Using a peristaltic pump, fill the hydrolysed tube with 1.0 M 4-(2,3-epoxy-propoxy)-benzaldehyde (25 ml) dissolved in dioxane, and recirculate the solution for 2 h at room temperature. Wash the tube alternately with dioxane and water.

9.2.3 *Coating of the Inside Tube Surface with Polyvinylalcohol forming IVa*

React tube III with an aqueous solution (2.2%, w/v; 0.5 M with respect to vinyl alcohol units) polyvinylalcohol (PVA; 20 ml) and conc. HCl (0.2 ml) for 15 h by recirculating the solution at 25°C. Wash the tube with water.

9.2.4 *Coating of the Inside of the Tube with PVA in the Presence of Terephthalaldehyde to form IVb*

React the tube III with an aqueous solution 2.2% (w/v) PVA (20 ml), with conc. HCl (0.2 ml), and add 0.1 M terephthalaldehyde dissolved in dioxane (1 ml). Incubate the tube for 30 min at 25°C by recirculating the mixture. Wash the tube with water and incubate with 0.1 M HCl for 15 h; wash the tube with water again.

9.2.5 *Activation of the Inside Tube Surface with 2,4,6-Trichloro-s-triazine forming Va and Vb (adapted from ref. 26).*

Incubate a 100 cm length of tube (IVa or IVb) with 20% (w/v) NaOH (10 ml) for

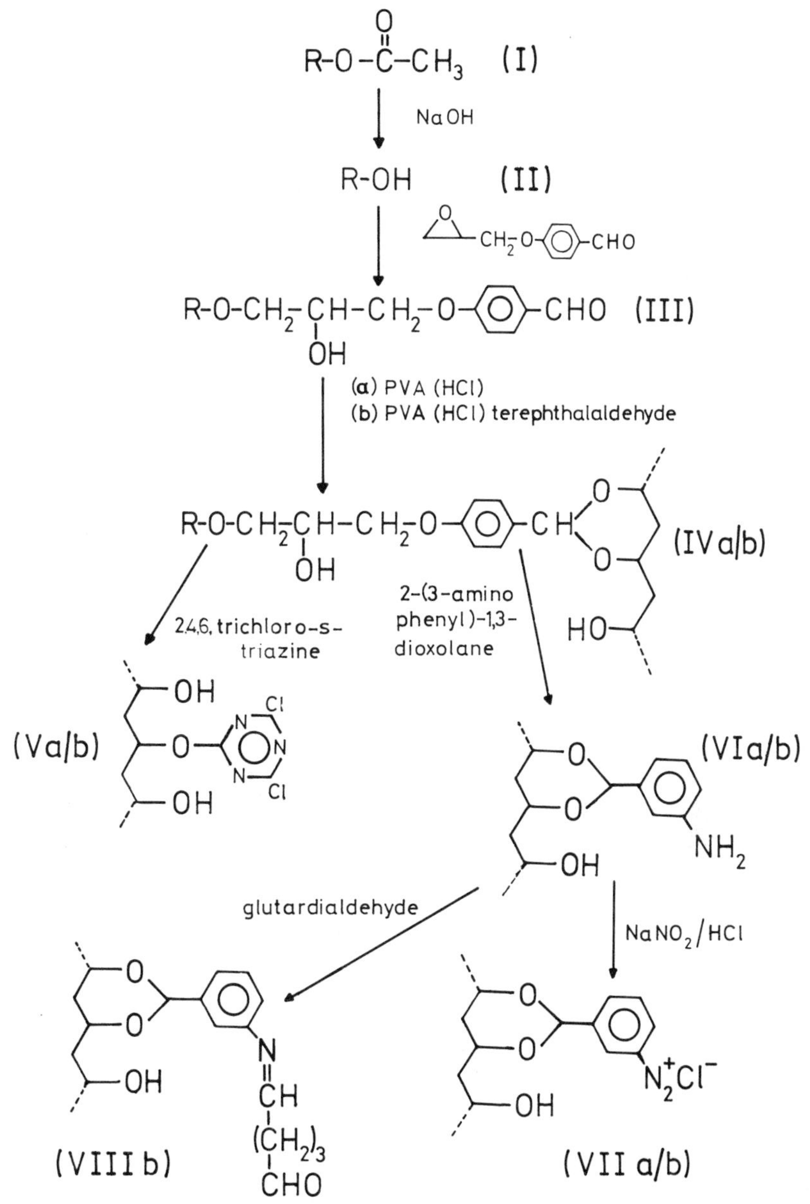

Figure 10. Structures of derivatives used to activate vinyl acetate-ethylene co-polymer tubes.

30 min at room temperature; remove the liquid under suction and add 10% (w/v) 2,4,6-trichloro-s-triazine (10 ml) in dioxane. After recirculating the solution through the tube for 10 min, stop the reaction by addition of 20% (v/v) acetic acid (10 ml). Wash the tube with dioxane and water.

17

9.2.6 *Derivatisation of Inside Tube Surfaces by Reaction with 2-(3-Amino-phenyl)-1,3-dioxolane, giving VIa and VIb*

React a 500 cm length of tube Ia or Ib with an aqueous solution of 2.2% (w/v) PVA, 1.0 M 2-(3-aminophenyl)-1,3-dioxolane (2 ml) in dioxane, and conc. HCl (0.5 ml). Recirculate the mixture at 40°C for 4 h; wash the tube with 0.01 M HCl and with water.

9.2.7 *Diazotisation to give VIIa and VIIb*

Recirculate a mixture of 0.5 M HCl (10 ml) and 1.0 M $NaNO_2$ (1 ml) at 0°C for 30 min through a l00 cm length of the tube VIa or VIb. Wash the tube with ice-cold water and the tube is ready for immobilisation of enzymes.

9.3 Immobilisation of Enzymes onto Activated Tube Surfaces

9.3.1 *Immobilisation of Trypsin on Triazine-activated Tubing*

Dissolve trypsin (5 mg) in phosphate buffer, pH 7.0 (I = 0.15)(5 ml), and couple to a 100 cm length of activated tube (Va or Vb) containing dichloro-*s*-triazinyl groups. After recirculating the enzyme solution for 3 min, stop the immobilisation reaction by addition of 1.0 M NH_3/NH_4Cl, pH 9.2 (3 ml). Wash the tube with phosphate buffer, pH 7.0, 0.5 M NaCl, and distilled water.

9.3.2 *Immobilisation of Trypsin by Azo-coupling on Tubes VIIa or VIIb*

Dissolve trypsin (5 mg) in phosphate buffer, pH 7.0 (I = 0.15) (5 ml), and couple at 0°C to a 100 cm length of tube (VIIa or VIIb). Recirculate the enzyme solution for 2 h, stop the reaction and remove the solution by suction. Wash the tube with phosphate buffer, pH 8.0, 0.5 M NaCl, and then distilled water. Collect the washing solution in a 100 ml measuring flask and determine the protein content with Folin's reagent, activate amount of immobilised enzyme. Fill the tube with water and store at 4°C.

9.3.3 *Immobilisation of Glucose Oxidase by Azo-coupling on Tubes VIIa or VIIb*

Incubate a 50 cm length of tube (VIIa or VIIb) for 1 h at 0°C with a solution of 2 mg glucose oxidase in 0.5 M triethanolamine buffer, pH 8.0 (4 ml). Perform the immobilisation reaction by recirculating the enzyme solution. Wash the immobilisation product with 0.5 M triethanolamine buffer, pH 8.0, 0.5 M NaCl, and distilled water.

9.3.4 *Immobilisation of Glucose Oxidase on Tube Vlb by the Glutaraldehyde Method*

Activate a 50 cm length of tube Vlb with a 5% (v/v) solution of glutaraldehyde in phosphate buffer, pH 8.0 (I = 0.15), by recirculating the solution at room temperature for 1 h, giving tube VIIIb. Wash the tube with water. Dissolve glucose oxidase (3 mg) in phosphate buffer (3 ml), pH 8.0 (I = 0.15), place in a 50 cm length of tube VIIIb, and recirculate the enzyme solution for 4 h at room temperature. Wash the tube with phosphate buffer, pH 8.0, 0.5 M NaCl and distilled water.

10. EUPERGIT C

Eupergit C (manufactured by Rohm Pharma, Darmstadt) consists of oxirane acrylic beads, obtained by co-polymerisation of methacrylamide, methylene bis-methacrylamide, glycidyl-methacrylate and/or allyl-glycidyl-ether. Eupergit C is claimed to be electroneutral and hydrophilic. The epoxide is chemically stable (pH 0 − 12) for several hours at room temperature in aqueous suspensions, but is stable when dry for months at − 30°C. Due to the chemical nature of the matrix-ligand linkage, ligands (such as proteins) are irreversibly immobilised to this matrix and no leakage of the ligand from the matrix occurs, within a range of pH 1 − 10.

The beads have an average diameter of 150 ± 20 μm (although microbeads for high performance chromatography are also available of 30 μm average diameter) with a macroporous structure which has a low water uptake. Under pressure Eupergit C compares favourably with agarose beads. No increase in resistance to flow is observed with Eupergit C on increasing the pressure from 0.1 to 1.0 bar. By contrast, agarose beads show an increased resistance to flow and clogging at 0.6 bar. Consequently Eupergit C is well suited even for downstream fixed-bed processes, allowing constant and high flow-rates for long periods of time.

The inner core of these oxirane beads is accessible to proteins with a molecular weight of less than 155 000 daltons. The beads have high binding capacity which reflects a high epoxide group content. Binding of 140 mg of albumin per gram of dry beads has been reported.

10.1 Reactions of Eupergit with Protein

10.1.1 *General Procedure*

(i) Protein preparations containing ammonium salts must be desalted by dialysis or ultrafiltration prior to binding; in the latter case, they can be simultaneously transferred to phosphate buffer. Dissolve the protein (10 − 200 mg) in 1.0 M potassium phosphate buffer, pH 7.4 (3.5 − 4.0 ml). Prepare this buffer by mixing 1.0 M KH_2PO_4 (23 ml) and 1.0 M K_2HPO_4 (77 ml) solutions, containing 0.5 mg/ml *p*-hydroxybenzoic acid ethyl ester (dissolved in isopropanol). Add the protein solution as uniformly as possible, using a pipette, to Eupergit C (1.0 g). This reaction is best carried out in a wide-necked flask having an internal diameter of some 15 − 20 mm; when the diameter of the flask is too small, optimal infiltration of the liquid into the beads cannot be guaranteed. Tightly stopper the flask and allow to stand for 16 − 72 h at room temperature (21 − 25°C). No stirring is required. Wash the gel on a sintered glass funnel (porosity 2 − 3) with distilled water (2 x 50 ml), then with 1.0 M NaCl (4 x 50 ml) and finally with the buffer (2 x 50 ml) in which it is intended to use the product. Include a preservative such as 0.5 mg/ml *p*-hydroxybenzoic acid ethyl ester.

(ii)   Removal of excess oxirane groups is generally unnecessary: if such treatment is required follow Section 10.2.

10.1.2 *Other Possibilities*

Oxirane groups can react with proteins over a wide pH range (pH $0-12$). Binding can take place in the acidic range (for example in HCl-NaCl for pH $0-3$ or acetic acid-acetate for pH $4-6$) as well as in the high alkaline range (e.g. in 0.2 M Na_2HPO_4). It is recommended that the pH at which optimum binding is achieved is determined for each individual protein. With many proteins, an ionic strength of 1.0 M potassium phosphate is optimal, although there are also cases in which lower ionic strengths are more successfully employed.

10.2 Formation of Eupergit C Derivatives with Low-molecular Weight Compounds

10.2.1 *Reactions with Mercaptans*

Carry out all these reactions in a fume cupboard.

(i)   *In aqueous media.* Pre-wash the oxirane beads (1 g) five times with water and filter under suction. Dissolve 2 mmol of the mercapto-compound (e.g., benzyl mercaptan) in water (2 ml) and adjust the pH value to $8-9$ with NaOH. Add the mercaptan solution to the beads, mix well by stirring and leave to stand overnight at 25°C. Wash the product six times with distilled water (10 vols). Use elemental analysis for sulphur to measure the degree of coupling.

(ii)   *In a non-aqueous medium.* Mix methyl alcohol [4 g, which need not be anhydrous; 10% (v/v) water actually serves to accelerate the reaction] containing Triton-B [5 mg, 40% (w/v) solution of benzyltrimethylammonium hydroxide in methanol] and benzyl mercaptan (180 mg) (~ 1.5 mmol), and add to Eupergit C (1 g). Shake the mixture for 72 h at room temperature (23°C), filter on a sintered glass funnel, and wash with methanol (five times the vols). Use sulphur analysis to measure coupling.

10.2.2 *Reactions with Hydroxyl Compounds, e.g., n-Hexanol*

Dissolve metallic sodium (250 mg) in *n*-hexanol (125 g) and add Eupergit C (25 g). Heat the suspension to 80°C and keep at this temperature for 48 h. Cool, and remove the excess hexanol by filtration; wash with methanol (5 x 100 ml) until the liquid is free of *n*-hexanol. Dry *in vacuo* for 48 h at 40°C. Analysis: determine the hexanoyl value by Zeisel's method. 20% of the oxirane groups are coupled to the hexanol derivative, whilst 80% oxirane groups are converted to the diol by alkaline hydrolysis.

10.2.3 *Reaction with Di-alkylamines, e.g., Hexamethylenediamine*

Dissolve hexamethylenediamine (800 mg) in distilled water (9 ml) and add Eupergit C (1 g). Gently shake the mixture for 2 h at room temperature. Wash with water (on a sintered glass funnel) until the washings give a neutral pH. Wash three times with ethanol and dry *in vacuo* for 72 h at room temperature. Analysis:

add HCl and determine chlorine quantitatively. This should show 100% conversion, i.e., introduction of 1 mol hexamethylenediamine/mol of oxirane groups.

10.2.4 *Removal of Excess Oxirane Groups (after Ligand Immobilisation)*

The content of oxirane groups is 800 − 1000 mol/g dry beads. This concentration of epoxide in Eupergit C, together with the macroporous nature of the matrix, frequently results in considerable protein-binding capacity. High activities of the immobilised product can also be achieved; the following immobilised activities have been reported:

500 Units/g of wet gel for penicillin-amidase (*Escherichia coli*)
600 Units/g for trypsin
500 Units/g for β-galactosidase
 90 Units/g for ribonuclease (bovine pancreas, RNA from yeast as substrate)

Activity units in all cases are equivalent to μmol substrate hydrolysed/min at 37°C.

With most enzymes, special matrix modification or removal of excess oxirane groups is unnecessary. However, the high content of oxirane groups is important for chemical modification of the matrix. One of the applications of the matrix which depends upon this feature is the use of phenylboronate-Eupergit. Haemoglobin A_{1c} binds to this matrix and the interaction depends on the high ligand concentration (CNBr activation of agarose does not provide adequate ligand concentrations) (F.A.Middle, unpublished observations).

In some cases it has been found that, after binding a protein, enough oxirane groups are left over to modify the matrix further, making it more anionic, cationic, hydrophilic or hydrophobic in character. This is achieved by allowing it to react with appropriate low molecular weight compounds (e.g., sulphydryl or amino compounds) carrying the functional groups desired. Thus, penicillin-amidase (*E. coli*) immobilised on Eupergit C can be considerably stabilised against alkaline deactivation by derivatisation with Cleland's reagent or other disulphydryl compounds.

With other ligands, however, such as antibodies or antigens, excess oxirane groups should be removed. For this reason alternative methods for the immobilisation process are recommended by the manufacturers. For details see special data sheets available from Rohm Pharma. An example is given below in Section 10.2.5.

10.2.5 *Method*

Treatment with mercaptoethanol removes the epoxide groups and keeps the matrix of the beads electroneutral and hydrophilic. Wash the wet beads containing bound protein on a sintered glass funnel (porosity 2) with 0.01 M phosphate buffer, pH 8.0 (10 vols). Add 5% (v/v) aqueous mercaptoethanol solution (adjusted to pH 7.6 − 8.0 with NaOH) per gram of wet beads and leave to stand overnight at room temperature. Wash the preparation with distilled water (10 x 5 vols) and transfer to the buffer in which it is intended to be used. Store as wet beads at 4°C. Prevent microbial growth by adding formaldehyde.

11. KIMAL – POROUS PARTICULATE ALUMINA

by R.C.Barabino and M.H.Keyes

11.1 Introduction

Studies of porous particulate alumina as an insoluble support for enzyme immobilisation began in 1974, largely as a result of flow-rate limitations imposed by polysaccharide supports when used in 'Partitioned Enzyme-Sensors' which were designed for clinical instrumentation. In the Owens-Illinois BUN instrument (27) for example, a sample containing urea flows through an immobilised urease column (3 mm x 76 mm) and is mixed into a second stream containing a strong base. The ammonia thus generated is detected by a gas electrode.

The flow-rate of this instrument is approximately 1 ml/min. If a polysaccharide support is used, column packing and erratic flow rates are frequently encountered. Polysaccharide supports are considered to be inadequate for column use because: (a) at the flow-rates typically required for rapid sample throughput in automated clinical instrumentation (>50 analyses/h), polysaccharide bead packing occurred and (b) disintegration of the polysaccharide beads occurred resulting in clogging, flow restriction and decreased detector lifetimes (28).

Consequently, a search for a porous inorganic support was initiated, with the objective of satisfying several requirements. The support should:

(i)   be physically difficult to deform or fracture under high flow-rates;
(ii)   be able to bind proteins in high yields and with activities approaching those obtained with many organic supports;
(iii) allow a wide selection of well-defined pore and particle sizes;
(iv)   possess long-term composite stability.

To satisfy these requirements, a porous particulate alumina, KIMAL was developed.

11.2 General Method for Preparation of Immobilised Composites

All reagents should be at least reagent grade and may be obtained from Aldrich Chemical Company, Fisher Scientific Company or Sigma Chemical Company.

(i)   Wash, by decantation, the porous, particulate alumina (KIMAL) with deionised water or 0.1 M HCl until free of fines, and activate in 6.0 M HCl for $1.5 - 3.0$ h under vacuum to remove trapped air in the pores.
(ii)   Dissolve the material to be bound in buffer, and store at 4°C until the alumina has been rinsed free of acid.
(iii) Decant the supernatant, cover the alumina with buffer and evacuate (water pump) the suspension and allow to equilibrate for $0.5 - 1.0$ h.
(iv)   Add the material to be bound and allow adsorption to proceed for $0.5 - 1.0$ h.
(v)   Add a solution of bifunctional cross-linking agent, either all at once or

continuously over 4 h. Stirring may be accomplished by means of an overhead stirrer.

(vi) Allow to react overnight at room temperature.

(vii) Wash the resulting product thoroughly, with (a) water and (b) buffer.

(viii) Store the suspension under buffer at $0-5°C$.

The amount of bound material can be determined by thermogravimetric analysis and is expressed as % weight loss following a heating step at 500°C.

11.3 Results and Discussion

Table 7 lists the characteristics of several support materials used to immobilise urease. The porous alumina developed at Owens-Illinois allows excellent enzyme loading, a reasonable range of flow-rates, excellent physical strength and good composite stability. Samples of urease immobilised onto alumina in columns for the Kimble Blood Urea Nitrogen Analyser typically gave accurate results (at 90% conversion) after 1000 injected samples (29).

Initial activities of immobilised urease on these alumina-based composites are well within the range of those obtained using polysaccharide supports, and the stability of the resulting inorganic composite and the measured flow-rates are excellent. As a result, rapid sample throughput may be readily achieved.

The development of a multi-component immobilised enzyme system for the determination of α-amylase in biological fluids (35) demonstrated the binding of substrates to KIMAL. In this system an immobilised starch composite, bound to acid-activated KIMAL in much the same manner as for enzymes, was used (36).

Table 7. The Characteristics of Several Support Materials Used to Immobilise Urease.

Matrix	Particle size (μm)	Activity (U/g)	Eluant flow at 0.5 ml/min	Physical strength	Reference
Sephadex					
G-25	100	1	good	good	30
G-200	80	115	fair	good	31
G-200	80	300	fair	good	30
Sepharose					
6B	180	500	fair	good	30
4B	80	1500	fair	good	30
Glass					
Bioglass 1000	75	15	good	good	32
Control pore (GPC 10)	175	20	poor	good	32
Silica gel					
Cab-O-sil (M7)	<40	70	poor	good	32
Polypore (8a-437)	<100	80	poor	fair	31
Alumina					
non-porous acidic	150	135	excellent	good	31
neutral	150	13	excellent	good	31
basic	150	58	excellent	good	31
KIMAL	325	450	excellent	good	31
	275	800	excellent	excellent	31
	160	1500	good	excellent	31

Table 8. Examples of Substrates Bound to KIMAL Alumina.

Starch	mg/starch/g KIMAL	Reference
Linthner (soluble)	0.010	33
Acid-activated amylose	0.112	31
Amylose	0.027	34
Amylose	0.045	34
Dextranase-activated amylose	0.099	31
Imidazole-activated amylose	0.015	31
CNBr-activated amylose	0.121	31
DEAE-dextran	0.500	34

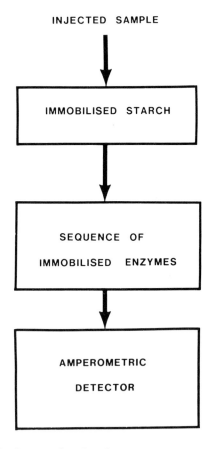

Figure 11. Flow diagram for the α-amylase detection system.

The amount of substrate bound was determined by iodine adsorption or thermogravimetric analysis. The amount of substrate thus bound to alumina is shown for selected samples in *Table 8*.

   Figure 11 summarises the flow stream for the detection of α-amylase by a clinical instrument. At reduced flow-rates four peaks were obtained. It is believed that these peaks may have been the result of isoenzyme separation on the substrate column. The single peak obtained at high flow-rate (>0.5 ml/min), was resolved into two further peaks at moderate flow-rate (>0.2 ml/min but <0.5 ml/min), and each of these became two well resolved peaks at low flow-rate (<0.2 ml/min), giving a total of four peaks. The literature concerning the elution chromatography of porcine α-amylase indicates that a mixture of α-amylase components may be separated electrophoretically into four components, two isozymes of pancreatic origin and two isozymes of salivary origin (37).

The unique characteristics of KIMAL developed in our laboratory enabled us to utilise alumina as an inorganic support not only for enzyme immobilisation, but also for use as an affinity chromatography medium.

12. CONTROLLED PORE GLASS

Controlled pore glass is the most commonly employed inorganic matrix for the immobilisation of biological molecules. Relatively few papers have reported the use of glass as an affinity chromatography matrix. Underivatised glass is known to adsorb enzymes non-specifically and glass is very slightly soluble at alkaline pH (>8.0). Glass is mechanically quite durable, is thermally stable and because it is an aerogel is chemically compatible with most organic solvents.

Controlled pore glass is synthesised by heating certain borosilicate glasses to $500-800°C$ for prolonged periods of time. These glass mixtures separate on such heat treatment into borate- and silicate-rich phases. The borate phase can be dissolved by treatment with acid, leaving a network of extremely small tunnels and pores ($25-70$ Å). The pore size distribution of controlled pore glass is extremely narrow giving the most uniform porosity of any material available.

Controlled pore glass consists of 96% silica and about 4% borate plus traces of other inorganic oxides. The surface of the glass contains most of the boron, up to 30% borate. The numerous boron Lewis acid sites will adsorb nucleophiles such as protein amines and/or adsorb ammonium ions to yield positively charged centres that lead to non-specific adsorption.

To prevent ionic adsorption of enzymes on glass surfaces, the glass is reacted with organic coatings that cover the silanol groups, thus minimising, if not eliminating, the ion-exchange character of glass. Derivatisation of controlled pore glass is most often performed by refluxing the beads in solutions of γ-aminopropyltriethoxysilane either in water [10% (w/v), pH 3.45] or toluene [10% (w/v)].

Other derivatives of controlled pore glass may be made by using triethoxypropyl glycidoxysilane as the alkylsilane (Pierce Chemical Company, Rockford, IL, USA).

13. DIOL-BONDED SILICA

by R.R.Walters

13.1 Introduction

Diol-bonded silica (38) has been used as the support material in most high-performance affinity chromatographic studies. Although such supports are com-

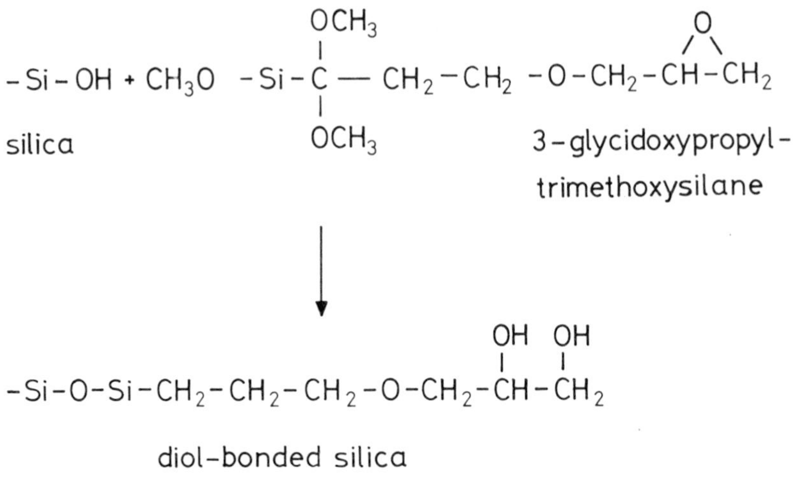

$$-Si-OH + CH_3O \; -Si-\overset{\overset{\displaystyle OCH_3}{|}}{\underset{\underset{\displaystyle OCH_3}{|}}{C}} - CH_2-CH_2 \; -O-CH_2-\overset{O}{\overset{/\backslash}{CH}}-CH_2$$

silica   3-glycidoxypropyl-

trimethoxysilane

$$\downarrow$$

$$-Si-O-Si-CH_2-CH_2-CH_2-O-CH_2-\overset{\overset{\displaystyle OH}{|}}{CH}-\overset{\overset{\displaystyle OH}{|}}{CH_2}$$

$$-Si-O-Si-CH_2-CH_2-CH_2-O-CH_2-\overset{\overset{\displaystyle OH\;\;OH}{|\;\;\;\;|}}{CH-CH_2}$$

diol-bonded silica

Figure 12. The reactions involved in the conversion of silica to diol-bonded silica.

mercially available, they can be easily synthesised from a silane and porous silica much more cheaply. The reaction is outlined in *Figure 12:*

Either anhydrous or aqueous procedures can be used. Anhydrous procedures produce a monolayer of silane, while aqueous procedures yield a polymeric layer. Aqueous methods are simple and may do a better job of covering the silanol groups which cause denaturation and non-specific adsorption. The polymerised layer may also be more resistant to hydrolytic cleavage (39).

Although many aqueous silanisation procedures can be found in the literature, we encountered difficulties when trying to synthesise supports of varying surface area. Both the concentration of silane and the number of molecules of silane relative to the surface area of the silica seem to influence the final surface concentration of the silane. These parameters cannot be held constant because of the wide variation in silica surface areas (see *Table 9*). Too little silane results in less than monolayer coverage of the silica, while too much silane results in agglomeration of the silica particles. Therefore, we developed empirical procedures which can be used for the silanisation of silicas over a wide range of surface areas.

13.2 Materials

(3-Glycidoxypropyl)trimethoxysilane is obtained from Petrarch Systems, 2731 Bartram Rd, Bristol, PA 19007, USA. Store the silane in a desiccator. LiChrosorb, LiChrospher and Spherisorb silicas are obtained from any one of several distributors. Sodium acetate and sulphuric acid can be reagent grade. All glassware must be carefully cleaned to remove any grease which could coat the silica.

13.3 Methods

(i)   Place the silica and the appropriate amount of 0.1 M sodium acetate buffer, pH 5.5 (*Table 9*) in a large test tube. The test tube should be large

Table 9. The Variation of the Degree of Silanisation for a Range of Porous Silicas.

Silica pore size, Å	Surface area, m^2/g^a	ml buffer/g silica	ml silane/g silica	% silane v/v	Fold silane[b]	µmol diol/g	Number of monolayers
60	500	15	0.75	5	3.0	910	0.8
80	220	10	0.50	5	4.0	720	1.3
100	250	10	0.50	5	4.0	500	0.9
300	250	10	0.50	5	4.0	520	0.9
500	50	4	0.10	2.5	4.0	190	1.7
1000	20	4	0.05	1.25	4.9	73	1.6
4000	6	5	0.05	1	16	33	2.4

[a]Manufacturer's data.
[b]Calculated amount of silane used relative to that needed for a monolayer.

	enough so that the silica will remain suspended while being gently shaken.
(ii)	Remove air trapped in the pores of the silica by sonicating the suspension and applying a vacuum with a water pump for 2 min.
(iii)	Add the appropriate amount of silane (*Table 9*) to the test tube.
(iv)	Stopper the tube and shake in a water bath at 90 ± 5°C for 5 h.
(v)	Thoroughly wash the silanised silica on a medium porosity sintered glass filter, first with water and then with water whose pH has been adjusted to pH 3.0 with H_2SO_4 (at this point the silica contains both epoxide and diol groups).
(vi)	Place the silica in a round-bottomed flask containing H_2SO_4, pH 3.0 (~ 50 ml/g silica).
(vii)	Add a few large glass beads, attach a condenser and heating mantle and reflux the mixture for 1 h to hydrolyse the remaining epoxide groups.
(viii)	Wash the diol-bonded silica with water, then with methanol, dry overnight at approximately 70°C, and store in a desiccator.

13.4 Discussion

A magnetic stirrer should not be used since the stirring bar may grind up the silica and adversely affect the chromatographic performance.

A minimum of 4 ml buffer/g silica are needed to suspend the silica. At least a 4-fold amount of silane should be used relative to a theoretical monolayer coverage of silane. The amount of silane to use can be calculated from the silica surface area provided by the manufacturer (m^2/g), the theoretical monolayer coverage of diol (2.3 µmol/m²) (40), and the silane molecular weight (236.3 g/ml) and density (1.070 g/mol). A larger excess of silane should be used when the silane concentration is less than 2% (v/v).

If desired, the diol content of the silica can be measured using the periodate oxidation method (41).

13.5 Determination of the Diol Content of Diol-bonded Silica by Periodate Oxidation

Diol groups can be quantitated according to the reactions outlined in *Figure 13*.

$$RCHOH.CH_2OH + HIO_4 \longrightarrow HCHO + HIO_3 + H_2O$$

$$HIO_3 + 5KI + 5H^+ \longrightarrow 3I_2 + 5K^+ + 3H_2O$$

$$HIO_4 + 7KI + 7H^+ \longrightarrow 4I_2 + 7K^+ + 4H_2O$$

$$I_2 + 2Na_2S_2O_3 \longrightarrow 2NaI + Na_2S_4O_6$$

Figure 13. The reactions used in the determination of diol groups in diol-bonded silica.

$$\mu mol \ diol = (ml \ blank - ml \ sample) \times M_{thio} \times$$

$$\frac{1 \ litre}{1000 \ ml} \times \frac{1 \ mol \ diol}{2 \ mol \ thiosulphate} \times \frac{10^6 \ \mu mol}{mol}$$

Figure 14. Calculation of diol concentration. The diol concentration of the sample can be obtained by dividing the μmol of diol by the sample weight.

The diol is oxidised with periodic acid. The iodine produced upon addition of potassium iodide is titrated with thiosulphate. A blank must also be run since both HIO_3 and excess HIO_4 produce iodine.

The modification of the periodate method (41) given here is suitable for up to 10 μmol of diol. The method can be scaled up or down by using proportionally larger or smaller volumes of all reagents.

13.5.1 *Materials*

Reagent grade chemicals are used for all solutions. A stock solution of 80% (v/v) acetic acid in water can be stored indefinitely, but the other solutions should be made up fresh each day: 10% (w/v) KI (5.0 g KI in 50 ml water); 0.02 M HIO_4 [0.25 g $HIO_4.2H_2O$ in 50 ml 80% (v/v) acetic acid]; and 0.02 M sodium thiosulphate (weigh exactly 4.9 g $Na_2S_2O_3.5H_2O$, mol. wt. = 248.18 or 3.1 g $Na_2S_2O_3$, mol. wt. = 158.11, and 0.04 g Na_2CO_3; dilute to exactly 1 litre with water).

13.5.2 *Methods*

(i)   Weigh a sample containing 1 − 8 μmol of diol into a 30 ml test tube. Add HIO_4 solution (0.5 ml).

(ii)   Sonicate this solution while applying a vacuum from an aspirator for 30 s to remove air from the pores of the silica.

(iii)   Allow the reaction to proceed at room temperature for 30 min, resuspending the silica intermittently.

(iv)   Add KI solution (0.25 ml) and titrate with thiosulphate (10 ml burette) until the last trace of yellow colour disappears. Near the end of the titration, rinse the walls of the test tube with water.

(v)   Titrate a blank solution containing only the HIO_4 and KI solution.

13.5.3 Calculations

First calculate the molarity of the thiosulphate solution, M_{thio}. The diol content can then be calculated (see *Figure 14*).

13.5.4 Discussion

Reagent grade thiosulphate is pure enough for the purpose of this assay. If better accuracy is desired it can be standardised with iodine (41) or primary standard thiosulphate can be prepared (42).

The volume of titrant needed for the sample should be less than 2/3 of the value of the blank; if the ratio reaches 3/4 all of the HIO_4 has been exhausted.

When the difference between volume of titrant used for the blank and sample is small, it is very important to run several replicates of the blank as well as the sample.

14. REFERENCES

1.   Nicolaisen,F.M., Meyland,I. and Schaumburg,K. (1980) *Acta Chem. Scand.*, **B34(2)**, 103.
2.   Porath,J. (1973) *Biochimie*, **55**, 943.
3.   Hjerten,S. (1964) *Biochim. Biophys. Acta*, **79**, 393.
4.   Determann,H., Rehner,H. and Wieland,T. (1968) *Makromol. Chem.*, **114**, 263.
5.   Determann,H. and Wieland,T. (1976) *Swedish Patent, No.* 382,066.
6.   Chitumbo,K. and Brown,W. (1971) *J.Polym. Sci.*, **C36**, 279.
7.   Edlund,O.H. and Andreassen,B.A. (1972) *German Offen.*, No. 2,138,905.
8.   Andreassen,B.A. (1972) *Swedish Patent*, No. 343,306.
9.   Andreassen,B.A. (1976) *Swedish Patent*, No. 382,329.
10.   Brown,W. and Chitumbo,K. (1972) *Chemica Scripta*, **2**, 88.
11.   Satake,T., Kano,I., Tsutsui,Y. and Yokota,K. (1974) *Japanese Kokai Patent*, No. 74 91,977.
12.   Satake,T. and Hata,K. (1975) *Japanese Kokai Patent*, No. 75 151,289.
13.   Chandler,B.V. and Johnson,R.L. (1975) *German Offen.*, No. 2,507,551.
14.   Chen,L.F. and Tsao,G.T. (1976) *Biotechnol. Bioeng.*, **18**, 1507.
15.   Peska,J., Stamberg,J. and Blace,Z. (1976) *Czech. Patent*, No. 172640, (1975) *German Offen* No. 2,523,893.
16.   Motozato,Y. (1978) *Japanese Kokai Patent*, No. 78 07,759 and 78 86,749.
17.   Kuga,S. (1980) *J.Chromatogr.* **195**, 221.
18.   Stamberg,J., Peska,J., Paul,D. and Philipp,B. (1979) *Acta Polymerica*, **30**, 734.
19.   Mislovicova,D., Gemeiner,P., Kuniak,L. and Zemek,J. (1980) *J.Chromatogr.*, **194**, 95.
20.   Hey,Y. (1983) Ph.D. Thesis, Liverpool University.
21.   Cruft,H.J. (1961) *Biochim. Biophys. Acta*, **54**, 609.
22.   Inman,J.K. and Dintzis,H.M. (1969) *Biochemistry (Wash.)*, **8**, 4074.
23.   Wichterle,O. and Lim,D. (1960) *Nature*, **185**, 117.
24.   Coupek,J., Labsky,J., Kalal,J., Turkova,J. and Valentova,O. (1977) *Biochim. Biophys. Acta*, **481**, 289.
25.   Manecke,G. and Vogt,H.-G. (1979) *J. Solid-phase Biochem.*, **4**, 233.
26.   Kay,G. and Crook,E.M. (1967) *Nature*, **216**, 514.
27.   Keyes, M.H. and Barabino, R.C.(1978) in *Enzyme Engineering*, Vol. 3, Pye, E.K. and Weetall, H.H. (eds.). Plenum Publishing Corporation, New York, p.51.
28.   Gray, D.N. and Keyes, M.H. (1977) *Chem. Tech.*, **7**, 642.
29.   Watson, B. and Keyes, M.H. (1976) *Anal. Lett.*, **9**, 713.
30.   Axen, R., Porath, J. and Ernback, S. (1967) *Nature*, **214**, 1302.

31. Keyes, M.H. (1976) *US Patent*, No. 3,933,589.
32. Richards,F.M. and Knowles,J.R. (1968) *J. Mol. Biol.*, **37**, 231.
33. Keyes,M.H. (1977) *US Patent*, No. 4001,085.
34. Caldwell,C.G., White,T.A., George,W.L. and Eberl,J.J. (1953) *US Patent*, No. 2,626,257.
35. Barabino,R.C., Gray,D.N. and Keyes,M.H. (1978) *Clin. Chem.*, **24**, 1393.
36. Barabino,R.C. and Keyes,M.H. (1980) *US Patent*, No. 4, 217, 415.
37. Meites,S. and Rogols,S. (1968) *Clin. Chem.*, **14**, 1176.
38. Regnier,F.E. and Noel,R. (1976) *J. Chromatogr. Sci.*, **14**, 316.
39. Watson,M.W. (1981) Ph.D. Dissertation, University of Minnesota.
40. Becker,N. and Unger,K.K. (1976) *Chromatographia*, **12**, 539.
41. Siggia,S. and Hanna,J.G. (1979) *Quantitative Organic Analysis*, published by Wiley, New York, p.42.
42. Wolf,A.A. (1982) *Anal. Chem.*, **54**, 2134.

APPENDIX I

Amicon Corporation, 17, Cherry Hill Drive, Danvers, MA 01923, USA or Upper Mill, Stonehouse, Gloucester GL10 2J, UK

Bio-Rad Laboratories, 2200 Wright Ave, Richmond, CA 94804, USA

Pierce Chemical Company, PO Box 117, Rockford, IL 61105, USA or 36, Clifton Rd., Cambridge CB1 4ZR, UK

Electro-nucleonics Inc., 12050 Tech Rd., Silver Springs, MA 20904, USA

L'Industrie Biologique Francaise, 101 Avenue de Verdun, 92390 Villeneuve la Garenne, France

Koch-Light-Genzyme, Hollands Rd., Haverhill, Suffolk CB9 8PU, UK

LKB-Produkter AB, PO Box 305, S-16126, Bromma, Sweden or 232, Addington Rd., South Croydon, Surrey CR2 8YD, UK

E.Merck GmbH, Frankfurter Strasse 250, D-6100 Darmstadt, FRG

Miles Laboratories Ltd., PO Box 37, Stoke Poges, Slough SL2 4LY, UK

Marine Colloids Inc., PO Box 308, Rockville, ME 04841, USA

Pharmacia - P.L. Biochemicals Inc., 1037, West McKinley Avenue, Milwaukee, IL, USA

Rohm Pharma GmbH, Postfach 4347, D-6100 Darmstadt, FRG

Serva Feinbiochemica GmbH & Co., D-6900 Heidelberg 1, FRG

Sigma Chemical Company, PO Box 14508, St Louis, MO 63178, USA or Fancy Rd., Poole, Dorset BH17 7NH, UK

CHAPTER 2

Activation Procedures

1. CYANOGEN BROMIDE

1.1 Attachment of Spacers to Matrices

Cyanogen bromide reacts with the vicinal diols of agarose, dextrans and cellulose to produce a reactive matrix that can be subsequently derivatised with either spacer molecules or ligands containing primary amines. This activation process has been up to now one of the most widely employed of all immobilisation methods (see *Figure 1*). However, two features of the above reaction have given rise to a vigorous search for alternatives: (i) because of the formation of charged isourea groups, CNBr activation invariably produces a bioselective adsorbent with anion-exchange properties, and (ii) the resulting isoureas are susceptible to nucleophilic attack and are therefore unstable.

The isourea formed on reaction with a primary amine is positively charged at physiological pH ($pK_a \sim 9.5$). If the alkyl group attached to the isourea function is markedly hydrophobic, the dielectric constant around the isourea groups will be very low, thus causing a large dispersal of the charge over the entire gel surface.

CNBr-activated matrices also display limited stability; hence coupling should be carried out immediately after activation. Significant leakage of coupled ligand from CNBr-activated supports also occurs, particularly in the presence of amine-containing buffers but, contrary to popular belief, Tris can be used since, although it contains such an amino group, it is extremely sterically hindered. For

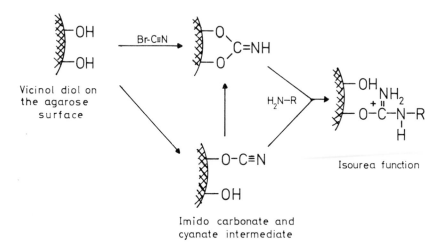

Figure 1. The cyanogen bromide activation of polysaccharides.

example, Tris will not react with ninhydrin and is difficult to aminoacylate under normal conditions (1).

Even with these limitations, cyanogen bromide-activated agarose has been widely and successfully employed in the synthesis of many affinity adsorbents. Because of the inherent dangers of cyanogen bromide, e.g., the release of HCN on acidification, the use of commercially pre-activated cyanogen bromide agarose is popular. For some purposes, however, it may be preferable to activate agarose in the laboratory just before use.

1.2 Titration Activation Method

CAUTION: CARRY OUT THIS REACTION ONLY IN A WELL-VENTI-LATED FUME HOOD

(i)   Add well-washed moist agarose beads (20 g) to distilled water (20 ml) in a 100 ml beaker equipped with a thermometer $(0-100°C)$, a pH meter and a 25 mm magnetic stirring bar (the bar should possess a central ridge to prevent destructive milling of the beads. A 1 mm band of Teflon or polyethylene will suffice if no ridged bar is available).

(ii)   Stir the suspension slowly, lower the temperature to $10-15°C$ by the addition of crushed ice and adjust the pH to 10.8 ± 0.1 by the addition of $1-2$ drops of 4.0 M NaOH.

(iii)   Initiate the activation procedure by the addition of CNBr and maintain the pH of the reaction at 10.8 ± 0.1 by manual titration with 4.0 M NaOH. The CNBr (100 mg/g moist weight gel) may be added as a crushed solid, or in acetonitrile (1 g/ml). The latter may be stored upright in a tightly stoppered vial (inside another screw-capped container at $-20°C$).

(iv)   Allow the temperature of the reaction to rise to $18-20°C$. The end-point is apparent when the addition of alkali is no longer necessary.

(v)   After the activation step is complete, cool the gel rapidly by the addition of crushed ice (beware: many ice machines use tap water and also introduce oil into the ice, test some before use and use only clean ice in this step) and pour into a large sintered glass funnel which has been pre-cooled with crushed ice.

(vi)   Filter the suspension rapidly into a Buchner flask (2 litres) containing solid ferrous sulphate to remove unreacted CNBr and cyanides as harmless ferrocyanide $(3-4$ M NaOH will also suffice to destroy CNBr and is suitable on a small scale).

(vii)   Wash the gel under suction with ice-cold distilled H_2O (1 litre) and buffer (1 litre) to be used in the coupling step.

1.3 The Coupling Step

Most unprotonated primary amines (note: Tris is an exception, see Section 1.1) couple efficiently to CNBr-activated agarose and dextran, to yield the structures shown in *Figure 1*. The formation of either N-substituted isourea derivatives or iminocarbonates depends on the support matrix used. The coupling pH should be chosen to be above the pK_a of the ligand but less than 10. Thus the coupling pH

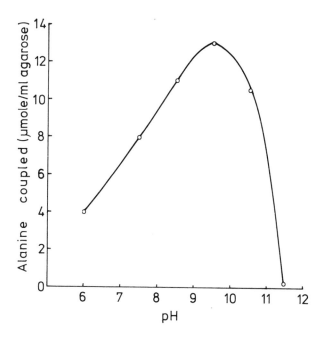

Figure 2. The effect of pH on the coupling of [¹⁴C]alanine to CNBr-activated agarose. Reproduced with permission from reference 2.

should be 7 – 8 for aromatic amines (pK_a ~5), 9.5 – 10 for amino acids (p$K_a\alpha$-NH$_2$ ~8) and 10 for aliphatic amines (pK_a ~9 – 10). Amine-containing buffers should be avoided during the coupling procedure [note: Tris can be used if required (1)], since these amino groups compete with the amino functions on the ligand for the activated groups. Borate or bicarbonate buffers are useful buffers. The coupling efficiency decreases at pH values above 9.5 – 10.0 (*Figure 2*) which reflects the sharp decline in stability of the activated complex. CNBr-activated Sepharose is also unstable at elevated temperatures and the coupling reaction is generally performed at 4°C.

1.4 An Alternative Activation Method after (3)

(i) Wash agarose (10 g) with phosphate buffer (25 ml) and filter to remove most surface water.

(ii) Add the agarose to a tube containing the volumes of buffer listed in *Table 1*.

(iii) Slowly add aqueous cyanogen bromide in the amounts listed, and shake the mixture gently for 10 min at 5°C.

(iv) When the reaction is finished, collect the product on a sintered glass funnel and wash with ice-cold distilled water until the filtrate is neutral.

(v) The activated agarose should be immediately coupled to the relevant ligand or spacer arm.

(vi) To couple the matrix to ligands or to spacers containing free primary amines, wash 20 ml of the CNBr-activated agarose (or the same amount of

Table 1. CNBr-activated Agarose.

Amount of derivatisation desired		Amount of buffer needed (ml)	Amount of aqueous CNBr to be used (ml)
Highly derivatised	4% gel	10	4
(~ 300 mmol/g)[a]	6% gel	15	6
Moderately derivatised	4% gel	0.8	0.8
(~ 140 mmol/g)[a]	6% gel	1.2	1.2
Weakly derivatised	4% gel	0.6	0.2
(~ 100 mmol/g)[a]	6% gel	0.7	0.3

[a]Cross-linked agarose (e.g., Sepharose CL) will yield only $25 - 30\%$ of the ligand concentration of uncross-linked agarose.

commercial CNBr-pre-activated agarose swollen according to the manufacturer's directions) with 0.25 M NaHCO$_3$ (50 ml).
(vii)   Suspend oxirane agarose (3 g) in 0.2 M Na$_2$CO$_3$, pH 11 (2 ml), containing $(50 - 500$ mg).
(viii)   Gently shake the mixture or stir at $0 - 4°$C overnight.

To ensure that all of the activated agarose hydroxyls are derivatised, unreacted cyanates may be blocked by shaking the beads in 1.0 M glycine, pH 8.0, for $6 - 12$ h at $0°$C.

2. OXIRANE-BASED METHODS

2.1 Introduction

Bis-oxiranes, such as 1,4-butanediol diglycidoxy ether (*Figure 3*), react readily with hydroxy- or amino-containing gels at alkaline pH to yield derivatives which possess a long-chain hydrophilic, reactive oxirane (epoxide), which, in turn, can be reacted with nucleophilic ligands (amines, phenols, etc.) to form bioselective adsorbents with a minimum number of hydrophobic and ionic groups. Oxirane-coupled ligands are widely used and extremely stable and the use of a long chain bis-oxirane reagent introduces a long hydrophilic spacer molecule between the ligand and the matrix backbone which may be desirable in certain applications.

Epoxide-activated Sepharose is available from Pharmacia, and oxirane acrylic beads (Eupergit C) from Rohm Pharma (see Chapter 1, *Table 3*).

2.2 Oxirane Activation of Agarose

(i)   Thoroughly wash beaded agarose (10 g) with distilled water and remove excess water using a sintered glass filter.
(ii)   Suspend the washed gel in 1.0 M NaOH (5 ml) containing 2 mg/ml NaBH$_4$ then add 1,4-butandioldiglycidoxyether and shake (or stir) the mixture gently for $5 - 10$ h at room temperature.
(iii)   Wash the activated gel thoroughly with distilled water and store at $4°$C.

Because loss of oxirane groups occurs slowly on storage, the activated gel should be used as quickly as possible. It has been reported that little or no loss of reactive functions occurs under these conditions for at least 1 week (4), although alkali accelerates decomposition of the oxirane.

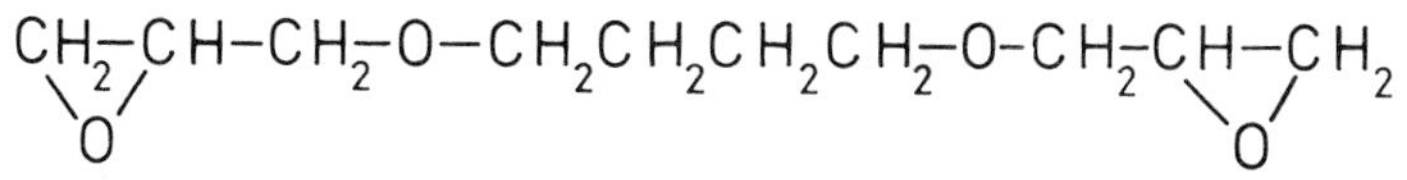

$$CH_2-CH-CH_2-O-CH_2CH_2CH_2CH_2-O-CH_2-CH-CH_2$$

Figure 3. 1,4-Butanediol diglycidoxy ether.

Gels with different oxirane concentrations may be prepared by varying either the reaction time or bis-oxirane concentration. Amines, hydroxyls, and other nucleophiles can be readily coupled to oxirane activated agarose.

2.2.1 Coupling to Oxirane Agarose

(i)   Suspend oxirane agarose (3 g) in 0.2 M Na_2CO_3 (2 ml), pH 11, containing amine (1.58 mmol).
(ii)   Keep the mixture at 4°C for 15 h and then wash with 1.0 M NaCl.
(iii)   Remove excess epoxide by storing the gel for 1 week in Na_2CO_3 buffer, pH 11 – 12, or by prolonged treatment with concentrated hydroxylamine at neutral pH. Alternatively, much briefer treatment of oxirane-agarose (4 – 6 h at pH 7.5) with mercaptans (2-mercaptoethanol, ethanedithiol, and thiourea) results in complete removal of the excess reactive epoxide.

The presence of cysteine, thiosulphate and thioglycollic acid should be avoided during the preparation since they will introduce ionic groups into the final product.

2.2.2 Recommended Procedure for Epoxy-activated Sepharose 6B

The following method is recommended by Pharmacia Ltd.

(i)   Weigh out the required amount of epoxy-activated Sepharose 6B.
(ii)   Wash and re-swell with distilled water on a sintered glass filter.
(iii)   Dissolve the ligand to be coupled in the coupling solution.
(iv)   Mix the ligand solution with the gel suspension for 16 h at 25 – 40°C, using a shaker in a water bath. Do not use a magnetic stirrer.
(v)   Wash away excess ligand using coupling solution followed by distilled water, 0.1 M bicarbonate buffer, pH 8.0, and finally 0.1 M acetate buffer, pH 4.0.
(vi)   Block excess groups with 1.0 M ethanolamine overnight.
(vii)   Wash away excess ethanolamine and store at 4 – 8°C.

2.2.3 Notes

(i) *Coupling solution*. Distilled water or aqueous buffers should be used whenever possible. Glycine or other nucleophilic buffers should not be used as these may couple to the oxirane groups. Carbonate, borate or phosphate buffers are useful trial buffers. Sodium hydroxide may be used for adjusting solutions to high pH values.

The ligand should be pre-dissolved in the coupling buffer. Certain organic solvents may be added to assist solution of the ligand. Dimethylformamide and dioxane may be used at up to 50% (v/v) of the final mixture. The pH of the aqueous phase should be adjusted after dissolving the ligand where necessary. The organic solvent usually causes a lowering of pH; pH paper should be used since organic solvents may damage electrodes.

(ii) *pH dependence.* Coupling is performed in the range pH 9 – 13. The stability of the ligand and the carbohydrate chains of the matrix, limit the maximum pH which can be used. The stability of the oxirane groups is also pH-dependent.

(iii) *Temperature dependence.* Coupling is performed in the range 20 – 45°C, preferably using a shaker in a water bath. Direct heating and magnetic stirrers should be avoided. If higher temperatures are used, coupling times may be decreased; the gels are mechanically stable at high temperatures and temperatures up to 100°C can be used for elution.

(iv) *Time of reaction.* The time of reaction depends greatly on the pH of the coupling solution, properties of the ligand and temperature of coupling. The efficiency of coupling is also pH- and temperature-dependent; 16 h at 20 – 45°C is most often used.

3. N-HYDROXYSUCCINYLAMIDOSUCCINYL-AGAROSE
by I.Matsumoto

Epoxy-activated agarose may be converted into amino derivatives using ammonia solutions. These derivatives of agarose may then be succinylated with succinic anhydride and activated with N-hydroxysuccinimide.

3.1 Amination of Epoxy-activated Agarose to Amino Agarose

(i)   Wash epoxy-activated agarose thoroughly on a sintered glass funnel with distilled water and then suspend in 0.880 ammonia solution (1.5 vols).

(ii) Incubate the suspension at 40°C for 1.5 h with shaking, then transfer to a sintered glass funnel and wash the gel with distilled water.

3.2 Preparation of the N-Hydroxysuccinimide Ester of Agarose

(i)   Wash amino agarose with 0.1 M NaCl and suspend in 0.1 M NaCl (1.5 vols).

(ii) Gradually add small portions of powdered succinic anhydride (0.08 g/g of moist amino agarose).

(iii) Maintain the suspension at pH 6.0 by the addition of 20% (w/v) NaOH.

(iv) When alkali consumption ceases, allow the suspension to stand for 5 h.

(v)   To remove labile carboxyl groups incubate the washed, succinylated agarose with 0.1 M NaOH for 30 min at room temperature.

3.3 Activation of Succinylamino-agarose with N-Hydroxysuccinimide

(i)   Wash the succinylamino agarose extensively with dioxane to make it anhydrous and suspend in dioxane (3 vols).

(ii) Add solid hydroxysuccinimide and dicyclohexylcarbodiimide to the suspension, to a final concentration of 0.1 M each, stirring continuously.

(iii) Incubate for 70 min, then wash the gel with dioxane (8 vols), followed by methanol (4 vols) (to remove the precipitated dicyclohexyl urea), and then with dioxane (3 vols).

3.4 Coupling of Lens culinaris Haemagglutinin to Activated Agarose

(i) Quickly wash moist activated Sepharose 4B (20 g) on a sintered glass funnel with water (50 ml) and then immediately transfer to a beaker containing lectin (100 mg) dissolved in 0.9% (w/v) NaCl – 0.01 M NaHCO$_3$, pH 7.5 (100 ml), containing 0.1 M methyl α-D-mannoside.

(ii) Incubate the suspension at room temperature for 1 h and allow to stand overnight at 4°C.

(iii) To block the remaining reactive groups, incubate the gel with 0.1 M triethanolamine HCl buffer, pH 9.0, for 1 h at room temperature.

4. TRIAZINE ACTIVATION OF AGAROSE

by C. Longstaff

(i)   Wash cross-linked agarose (1 vol) on a sintered glass funnel with distilled water (5 vols), followed by 12% (w/v) NaOH (5 vols).

(ii) When the NaOH has drained under gravity, add the gel to 12% (w/v) NaOH (1 vol).

(iii) Cool the suspension in an ice/salt bath to +5°C (±1°C).

(iv) Dissolve trichloro-*s*-triazine (recrystallised once from boiling chloroform to remove cyanuric acid) in acetone, to give a 0.5 M solution, and cool in ice/salt to approximately −5°C.

(v)   Pour 2 volumes of this solution carefully onto the gel suspension in such a way as to form a bilayer.

(vi) Surround the mixture with ice/salt and stir slowly using a ridged magnetic stirrer. The two layers gradually mix over the next 100 min and the bilayer eventually disappears.

(vii) During this period, monitor the temperature at the centre of the mixture and cool the reaction vessel in ice.

(viii) Once the temperature has stabilised, the ice may be removed and the mixing continued at room temperature.

(ix) Wash the activated gel with 50% (v/v) acetone (5 vols) followed by water (5 vols).

Triazine-activated gels can be stored for several weeks in distilled water at 4°C without appreciable loss of activity. Coupling is carried out at moderately alkaline pH, e.g., 0.5 M potassium phosphate buffer pH 7.8.

4.1 Notes

The extent of activation may be decreased by using a higher concentration of NaOH [up to 20% (w/v)] (*Table 2*) or a lower concentration of trichloro-*s*-triazine (*Table 3*).

Table 2. Effect of NaOH Concentration on Triazine Activation (Estimated from the Amount of Hexanediamine Coupled to Activated Gel) Determined with Ninhydrin.

NaOH conc. % (w/v)	Time (min)	Hexanediamine coupled ($\mu mol/mg/gel$)
10	90	0.92
15	130	0.68
17	130	0.64
20	120	0.46

Table 3. Influence of Trichloro-s-triazine Concentration on Triazine Activation of Agarose (Estimated from the Amount of Ethylenediamine Coupled to Activated Gel) Determined with Ninhydrin.

Trichloro-s-triazine concentration (mol/l)	Ethylenediamine coupled ($\mu mol/mg$ gel)
0.3	0.23
0.15	0.17
0.075	0.04

4.2 Coupling of Ethylenediamine or Aliphatic Diamine to Triazine-activated Agarose

Suspend moist activated gel (2 g) in water (2 ml). Add 1 M ethylenediamine (4 ml) and mix gently for a minimum of 2 h, or overnight at room temperature. The degree of activation can be measured using the ninhydrin reagent.

4.3 Triazine Activation of Filter Paper

(i)   Soak filter paper (20 x 20 cm) in 3.0 M NaOH for 1.5 h.

(ii)   Remove paper carefully and allow excess NaOH to drain away.

(iii) Lay the paper in a tank and add a 0.5 M solution of trichloro-s-triazine in acetone (1 litre).

(iv)   Leave for 1 – 3 h with occasional mixing. During the activation a precipitate forms and the paper may become distorted.

(v)   Wash the activated paper with 50% (v/v) acetone (1 litre) for 15 min, and then water (1 litre) for 15 min.

4.4 Coupling of m-Aminophenylboronic Acid to Triazine-activated Paper

(i)   Immerse the washed, activated paper in 0.5 M potassium phosphate buffer, pH 7.8, containing *m*-aminophenylboronic acid (1 g/litre). Leave for 24 h.

(ii)   Remove excess reactive groups by immersing the coupled paper in 0.5 M ethanolamine for 24 h.

(iii) Wash paper with 5% (v/v) acetic acid and 1.0 M KCl, and leave to dry overnight.

5. 2,4,6-TRIFLUORO-5-CHLOROPYRIMIDINE ACTIVATION OF AGAROSE

by T.C.J.Gribnau

5.1 Procedures

Reactions are performed in a 100 ml three-necked round-bottomed flask with a cooling mantle (*Figure 4*). The interior of the flask should be previously siliconised: rinse the flask after thorough cleaning with a 5% (w/v) solution of Dow-Corning 200 Fluid 350 CS in chloroform; after evaporation of the chloroform, heat for 30 min at 300°C. Suspend the agarose in the reaction medium by means of a 'Vibro-Mixer' (Chemap A.G., Männedorf-Zürich, Switzerland).

5.1.1 *Method A*

(i)   Suspend Sepharose 4B (25 g, filtered gel) in 3.0 M NaOH (50 ml) and rotate slowly for 15 min at room temperature. Filter off this 'alkali-Sepharose' on a sintered glass funnel and remove the liquid by suction (water pump).

(ii)   Suspend this material in 40 ml xylene-dioxane (1:1, w/w), agitate the suspension with a Vibro-Mixer at maximum frequency, and cool to 0°C.

(iii)   Add dropwise a solution of 2,4,6-trifluoro-5-chloropyrimidine (FCP) (4 – 5 g, 2.5 – 3.0 ml) in the same xylene-dioxane mixture (4 ml) at such a rate that the temperature does not rise above 0.5°C; this procedure requires about 1 h.

(iv)   Quench the reaction by addition of concentrated acetic acid (4.2 ml).

(v)   Filter off the activated agarose on a sintered glass funnel and wash with xylene-dioxane (200 ml), dioxane (200 ml), and finally water (1 litre). Store the gel at 4°C.

In contrast to native Sepharose 4B, FCP-agarose does not dissolve in water at temperatures above 40°C. Wash an aliquot with acetone, remove fluids by suction (water pump), and dry *in vacuo* over concentrated sulphuric acid. The N, F,

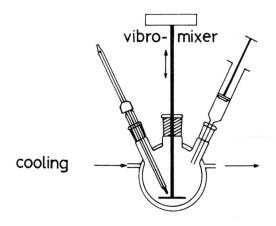

Figure 4. Apparatus for the FCP activation of agarose.

and Cl contents are determined by elemental analysis. Some representative examples are given in *Table 4*.

5.1.2 *Method B*

(i)   Pack 'alkali-Sepharose' 4B ($\sim$25 g, filtered gel) prepared as in Method A, into a plastic syringe (20 ml), the needle of which has been replaced by a Teflon tube (length 7 cm, internal diameter 3 mm), as in *Figure 4*.

(ii)   Add the Sepharose to a solution of FCP ($4-5$ g) in xylene-dioxane (1:1 w/w) (40 ml), previously cooled to 0°C, at such a rate that the temperature does not rise above 4°C; this requires about 25 min.

(iii)   Allow the reaction to continue for another 15 min at 0°C, and add concentrated acetic acid (4.2 ml) at the end of this period.

(iv)   Isolate the FCP-agarose as described in Method A.

Representative examples of elemental analyses are given in *Table 5*.

5.1.3 *Method C*

(i)   Wash Sepharose 4B ($\sim$25 g, filtered gel) with NaOH or Na_2CO_3 (250 ml) (see *Table 6*) and remove fluids on a sintered glass funnel (water pump).

(ii)   Pack the 'alkali-Sepharose' into a plastic syringe, as described in Method B.

(iii)   Inject the total amount of alkali-Sepharose in one portion into a solution of FCP in 40 ml xylene dioxane (1:1 w/w), previously cooled to -10°C. The temperature rises to $+9$°C, but then drops quite rapidly.

(iv)   Allow the reaction to proceed for 15 min at $+3$°C, then add a slight excess of acetic acid ($\sim$4.2 ml).

(v)   Isolate the activated agarose as described in Method A.

Table 4. Representative Elemental Analyses for FCP-activated Agarose Beads Prepared by Method A.

Batch	%N	%F	%Cl
a	4.6	2.9	5.9
b	4.7	2.8	5.7
c	4.6	2.9	6.0

Table 5. Representative Elemental Analyses for FCP-activated Agarose Beads Prepared by Method B.

Batch	%N	%F	%Cl
d	5.8	5.3	6.5
e	5.7	5.4	6.6
f	5.7	5.05	6.4

Table 6. Representative Elemental Analyses for FCP-activated Agarose Beads Prepared by Method C.

Batch	Alkali	FCP (g)	% M	% F	% Cl
g	3.00 M NaOH	1.07	0.8	0.6	1.0
h	0.27 M NaOH	1.07	1.7	1.3	1.9
i	0.10 M NaOH	1.00	0.9	0.9	1.15
j	0.27 M Na$_2$CO$_3$ (pH 11.4)	1.03	0.6	0.5	0.55
k	1.32 M NaOH	5.06	4.2	4.0	4.7

5.2 Coupling of FCP-agarose with Hexamethylenediamine

(i) Take FCP-activated Sepharose 4B, remove fluids with a sintered glass funnel by suction (water pump), suspend the gel in an aqueous solution of hexamethylenediamine (HMDA) and rotate the suspension slowly for 15 – 20 h at room temperature.

(ii) Filter off the product, wash thoroughly with double-distilled water, 0.05 M HCl (twice the gel volume), double-distilled water, 0.1 M NaOH (five times the gel volume) and again with double-distilled water until the eluate is neutral. A qualitative check on the presence of primary amino groups may be performed by testing with ninhydrin reagent.

(iii) Wash an aliquot of the HMDA-Sepharose with methanol, remove fluids by suction and dry the gel *in vacuo*.

The N,F and Cl contents are determined by elemental analysis.

5.3 Coupling of FCP-agarose with Aniline

(i) Activate Sepharose 4B (23 g, filtered gel) as described in Section 5.1 (Method A).

(ii) Suspend the activated product in a solution of aniline (1.74 g) in 0.07 M phosphate buffer (pH 7.0) (25 ml) and peroxide-free dioxane (10 ml).

(iii) Rotate the suspension for 26 h at room temperature.

(iv) Filter off the product, wash with water-dioxane (2.5:1 v/v) and water.

(v) Wash an aliquot with acetone, remove fluids by suction and dry *in vacuo*: N, F and Cl contents are determined by elemental analysis.

Isolation of the product is performed as described for HMDA-agarose.

5.4 Coupling of FCP-agarose with Imidazole

(i) Wash FCP-Sepharose 4B (5 g, filtered gel) with 0.2 M borate buffer, pH 9.0 (50 ml), suspend in 0.5 M imidazole (5 ml) in the same buffer, and rotate the suspension for 17 h at room temperature.

(ii) Filter off the product, wash with water, 0.1 M HCl, water, 0.1 M NaOH, and double-distilled water.

(iii) Wash an aliquot with methanol, remove fluids by suction, and dry *in vacuo*.

N, F and Cl contents are determined by elemental analysis.

5.5 **Coupling of FCP-agarose with Ethanethiol**

(i)   Suspend FCP-Sepharose 4B (2 g moist weight) in a mixture of ethanethiol (1 ml), 0.2 M borate buffer, pH 9.2 (5 ml), and ethanol (7 ml).

(ii) Rotate the suspension for 21 h at room temperature.

(iii) Filter and wash the product thoroughly with ethanol-water (1:1; v/v), ethanol and finally acetone.

5.6 **Coupling of FCP-agarose with Glycylglycine/p-Aminobenzamidine (Biospecific Adsorbent for Trypsin)**

(i)   Suspend FCP-Sepharose 4B (11 g filtered gel; batch j, see *Table 6*) in a solution of glycylglycine (0.74 g) in 5 ml of buffer, pH 10.6 (0.5 M Na_2CO_3/ $NaHCO_3$).

(ii)   Rotate the suspension for 27 h at room temperature.

(iii) Filter the product, wash with double-distilled water (250 ml).

(iv)   Suspend the filtered gel in a solution of *p*-aminobenzamidine dihydrochloride (1.25 g, 6 mmol) in double-distilled water.

(v)   Adjust the pH of the suspension to 5.0 and add 1-(3-dimethylamino-propyl)-3-ethylcarbodiimide hydrochloride (1.15 g, 6 mmol).

(vi)   Allow the reaction to proceed for 2 h at pH 5.0 and room temperature.

(vii) Filter off the product and wash with double-distilled water, 0.5 M NaCl (100 ml), double-distilled water (250 ml).

(viii) Store the gel in 0.05% (w/v) NaN_3 at 4°C.

The ligand content of the adsorbent was 108 μmol/g dry gel material, as calculated from elemental analysis.

6. CARBONYLDIIMIDAZOLE ACTIVATION

Polysaccharide matrices may be activated using carbonylating agents of the general formula:

$$O = \overset{\overset{\displaystyle Y}{|}}{C} - X$$

where Y is a good leaving group readily displaced by an amine, and X is either the same as Y or another appropriate leaving group which combines with a hydrogen atom during carbonylation. Suitable carbonylating reagents include N,N'-carbonyl diimidazole (CDI); N,N'-carbonyl di-1,2,3-benzotriazole (CBT); or N,N'-carbonyl di-1,2,4-triazole (CDT).

6.1 **Activation of Sepharose CL 6B with Carbonyldiimidazole**

(i)   Wash Sepharose CL 6B (3 g moist weight) sequentially with water, dioxane-water (3:7 v/v), dioxane-water (7:3 v/v) and dioxane (50 ml of each) and suspend in dioxane (5 ml).

(ii) Add CDI (0.12 g). The reaction between Sepharose and CDI can be left until it is convenient to continue the washing and coupling steps, the yields being identical for reaction times between 0.25 and 6 h. This is in direct contrast to the CNBr method which requires coupling to be done as quickly as possible. In addition, the activation can be greatly increased if more CDI is used.

(iii)   After activation, wash with dioxane (100 ml) and use immediately.

6.2 Coupling of Ligands to Activated Matrices

Treat the activated matrix overnight with *n*-butylamine (1.4 g) in water (9 ml) at pH 10 and wash sequentially with water (200 ml), 1.0 M NaCl (100 ml) and water (200 ml). On the grounds of cost and capacity, the CDT and CDB activated matrices do not appear to offer any advantages over the cheaper CDI method except that the CDT method offers a greater rate of coupling. The CDI method seems to be comparable with the standard CNBr method with the added advantage of being more pleasant to handle.

One of the greatest advantages of the methods based on the carbonyl reagents over the CNBr method lies in the fact that no charged groups are introduced into the matrix during the coupling step.

6.3 Recommended Procedure for the Use of Commercially Available CDI-activated Agarose

CDI-activated 6% cross-linked agarose is available from Pierce Chemical Company as Reacti-Gel (6X). Reacti-Gel (6X) is 6% cross-linked beaded agarose which has been derivatised with CDI to give a highly activated imidazolyl-carbamate matrix.

6.3.1 Coupling Procedure

Coupling with proteins is primarily through accessible α- and ε-amino groups. Proteins can be coupled directly through multi-point attachment sites or by use of a spacer. Proteins not compatible with (or sensitive to) pH values of 9 or greater can be coupled at pH 8.5 in 0.1 M borate buffer. If this coupling condition is not appropriate for the protein, coupling can be effected via a spacer by using a water-soluble carbodiimide at pH 4.5 – 5.5 for 24 h at room temperature.

(i)   Remove excess acetone from Reacti-Gel (6X) on a sintered glass funnel using gentle suction. The moist cake should no longer drip acetone, but do not dry gel completely.

(ii)   Break up the Reacti-Gel (6X) cake into finely divided pieces.

(iii)   Add the gel directly to the coupling buffer containing ligand.

(iv)   For proteins that are sensitive to small amounts of acetone, wash the gel free of the solvent before use. This is accomplished by taking the desired amount of gel and washing quickly with several bed volumes of ice cold water.

Bovine thyroglobulin, bovine thyroid stimulating hormone, porcine insulin and human immunoglobulin can be successfully coupled to Reacti-Gel (6X) in 87 – 100% yield and have been shown to retain their biological activity under these coupling conditions.

CDI-activated cross-linked beaded Dextran Reacti-Gel (25DF) is also available commercially (from Pierce Chemical Company).

7. MIXED ANHYDRIDE ACTIVATION PROCEDURES

The procedure given below is first applied to the activation of a ligand rather than the matrix (Section 7.1). In Section 7.2 the chemistry has been adapted to carboxylated matrices but only applies to those which can be dehydrated (e.g., carboxy-Eupergit C which may be prepared by treatment of Eupergit C with glycine).

7.1 Preparation of 17 β-oestradiol-6-(O-carboxymethyl) Oxime BSA

(i)   Dissolve 17 β-oestradiol-6-(O-carboxymethyl) oxime (180 mg) and tri-*n*-butylamine (0.22 ml) in dry dioxane (7.5 ml) and cool in ice.

(ii)   To this mixture add isobutylchlorocarbonate (55 μl) and leave the mixture for 30 min, keeping the temperature below 10°C.

(iii)   Add a solution of bovine serum albumin (BSA, 0.58 g) in a mixture of water (15.3 ml), dioxane (10.3 ml) and 1.0 M NaOH (0.58 ml); allow to react for 4 h at $5 - 10°C$ maintaining the pH at 8.0 with NaOH.

(iv)   Remove low molecular weight materials from the reaction mixture using a Sephadex G-25 column (30 x 2.5 cm) pre-equilibrated in 0.9% (w/v) NaCl solution, by eluting with 0.9% (w/v) NaCl solution.

(v)   Acidify the purified conjugate with dilute HCl to pH 4.6, and leave for 4 days at 4°C.

(vi)   Recover the precipitate which forms by centrifugation, resuspend in a small volume of water and dissolve by adding enough solid $NaHCO_3$.

(vii)   Lyophilise the solution.

7.2 Preparation of Isochlorocarbonate-activated Matrices

(i)   Dry the carboxylated matrix over P_2O_5.

(ii)   Weigh out 1 g in a dry pre-weighed vial.

(iii)   Suspend the matrix in dry dioxane (5 ml) and add tri-*n*-butylamine (1 ml).

(iv)   Allow to stand for 10 min and wash the matrix with dry dioxane (10 ml).

(v)   Cool to 0°C, add isobutylchlorocarbonate (2 ml) and leave for 30 min, keeping the temperature below 10°C.

(vi)   Rapidly wash the matrix with dry dioxane (20 ml), dioxane:water (1:1, 50 ml) and water (50 ml).

(vii)   Add a solution of the amino compound to be coupled $(50 - 200 \ \mu mol$ of amino groups) dissolved in buffer at pH 8.0. Phosphate-NaOH or borate buffer $(0.02 - 0.2$ M) can be used.

(viii) Wash the matrix free from uncoupled ligand and determine the ligand concentration either by measuring the uncoupled ligand or by assaying the matrix.

8. PERIODATE OXIDATION

Polysaccharide matrices may be activated by the oxidation of the vicinal diol groups using sodium meta-periodate ($NaIO_4$), which oxidises these sugar diols to generate aldehyde groups. The method represents a rapid, convenient alternative to the use of cyanogen bromide. Sodium meta-periodate is also soluble in aqueous media and may therefore be readily removed after reaction by washing. The activated gel can be stored at 4°C for many days without appreciable loss of activity. The aldehyde groups formed react with most primary amines between pH values 4 to 6 to form Schiff's bases, which may be subsequently stabilised by reduction with borohydride (NaBH$_4$) or preferably cyanoborohydride (NaBH$_3$CN).

(i)   Mix the gel with an equal volume of 0.2 M NaIO$_4$.

(ii)   Place in a tightly closed polyethylene bottle and gently shake on a mechanical shaker for 2 h at room temperature.

(iii)   Filter the periodate-oxidised matrix, and wash thoroughly on a sintered glass funnel with distilled water.

8.1 Preparation of Aminohexyl-agarose

8.1.1 *Coupling Method A*

(i)   Add periodate-oxidised agarose (100 ml) to 2.0 M aqueous hexanediamine (100 ml) at pH 5.0 and gently shake for 6 – 10 h at 20°C.

(ii)   Raise the pH to 9.0 by the addition of solid Na$_2$CO$_3$, then add freshly prepared aqueous 5 M NaBH$_4$ (10 ml) in small portions at 4°C with gentle stirring over a period of 12 h.

(iii)   Wash the reduced gel with 1.0 M NaCl until the filtrate contains no amine.

8.1.2 *Coupling Method B*

(i)   Mix together equal volumes of periodate-oxidised agarose, previously washed with 0.5 M phosphate buffer (pH 6.0), with 25 mM amine-containing ligand in 0.5 M phosphate buffer (pH 6.0) and 0.5 mM NaBH$_3$CN and leave for 3 days at room temperature.

(ii)   Wash the gel and reduce remaining aldehyde groups with an equal volume of 1.0 M NaBH$_4$ for 15 h at 4°C.

(iii)   Wash the gel extensively with distilled water before use.

8.2 Periodate Activation of Cellulose

(i)   Using either a rotary or end-over-end type mixer, mix beaded or fibrous cellulose (10 g) with 0.5 M sodium meta-periodate (1 litre) at room temperature for 2 h; then add ethylene glycol (30 ml) and continue mixing for an additional hour.

(ii)   Filter the cellulose and wash with water (1 litre).

(iii)   Suspend the washed gel in an equal volume of water.

(iv)   To this slurry add approximately 30 mmol of adipoyl dihydrazide, dissolved in a minimum of water, and adjust the pH of the resulting mixture to pH 5.0 with HCl.

(v)   After 1 h of further agitation, adjust to pH 9.0 with solid sodium carbonate.

(vi)   Add sodium borohydride (1 g), stir and leave to react for 12 h in an open vessel at 4°C (CAUTION; hydrogen gas is evolved).

(vii) Filter the cellulose and wash with 0.1 M acetic acid (200 ml), followed by distilled water until the filtrate is neutral.

8.3 Preparation of Immunoadsorbents with Very Low Non-specific Binding Properties using Periodate-oxidised Cross-linked Sepharose

by E.A.Fischer

In the immunoadsorption of serum components, non-specific binding to the matrix is often a serious problem which not only lowers the degree of purification but also the yield obtained. This effect is particularly pronounced with cyanogen bromide-activated agarose. Certain classes of IgG bind tightly to such gels. This binding is independent of the procedure used for the inactivation of unreacted groups on the gel. Comparing differently derivatised carbohydrate matrices, it has been observed that activation procedures which introduce charged groups into the matrix generate non-specific binding characteristics (5). Cyanogen bromide activation leads to the formation of charged isoureas and shows undesirable properties. Considerable improvement is found with Affi-Gel type matrices where activated carboxylic acid esters are attached to the agarose matrix via ether linkages. One drawback of these matrices, however, is that the active groups are easily hydrolysed, which can lead to very poor coupling yields. Periodate-activated Sephadex (G50 or G75) has been described (6). After coupling the immunoglobulin and reduction of Schiff's bases, the resulting immunoadsorbents showed very low non-specific binding, but the gels were mechanically very soft and tended to clog. Additionally, because antibodies were excluded from the inner space of the beads, the capacities of these gels were very low. Although agarose has a larger pore size, this gel is not efficiently activated by periodate oxidation, as the reaction is limited to the chain ends of the polysaccharide molecules. It is, however, possible to activate cross-linked agarose by periodate oxidation. In a side reaction of the cross-linking process with epichlorohydrin, glycols are generated on the gel. Periodate oxidation of these glycols produces aldehydes, which can be reacted with protein amino groups. The Schiff's bases formed can be subsequently reduced as with other matrices. Immunoadsorbents prepared in this way have excellent chromatographic properties, combining the rigidity of agarose beads with chemical stability and low non-specific binding properties. High flow rates can be obtained even with chaotropic agents like 6.0 M guanidine-HCl (7).

8.4 Coupling of Antibodies and Antigens to Periodate-oxidised Sepharose CL 6B

8.4.1 *Pretreatment of Gel*

(i)   Wash Sepharose CL 6B on a sintered glass funnel and wash with distilled water (5 – 10 volumes).

(ii)   Remove the bulk of interstitial water.

(iii) Place 35 g of the Sepharose in a 200 ml polyethylene bottle.

(iv)   Add 0.07 M sodium periodate solution (70 ml) in water and slowly rotate the bottle for 45 min at room temperature.

(v)   Add 2.0 M ethylene glycol solution (35 ml) and rotate the bottle for another 30 min.

(vi)   Filter the gel on a sintered glass funnel and wash with water followed by 0.01 M sodium carbonate buffer, pH 9.5.

(vii)   Weigh appropriate quantities into the relevant protein solutions.

8.4.2 *Preparation of Protein Solutions for Coupling*

Prepare protein solutions in phosphate buffered saline (PBS) containing 20 mM phosphate and 0.13 M sodium chloride, pH 7.4. Make the solution 0.01 M in sodium carbonate (with 0.2 M) and adjust to pH 9.5 with sodium hydroxide. Protein concentrations should be between 5 and 30 mg/ml.

8.4.3 *Coupling of Proteins to Activated Gel*

(i)   Add activated Sepharose (1 – 1.5 g) to the protein solution (1 ml) prepared as above, ensuring that the resultant slurry is still mobile.

(ii)   Allow the coupling reaction to proceed for 48 h, slowly rotating the bottle at 4 – 8°C.

(iii)   Remove unreacted protein from the gel by filtration and washing with ice-cold PBS.

(iv)   Remove interstitial liquid.

(v)   Reduce Schiff's bases and residual aldehyde groups by adding freshly prepared ice-cold sodium borohydride solution (1 mg/ml in PBS) (2 ml/g gel).

(vi)   Allow to react by slowly rotating the bottle for 1 h in the cold room.

(vii)   Wash the gel with PBS, followed by the chaotropic agent to be used in the immunoadsorption experiment.

(viii)   Wash exhaustively with PBS and the gel is ready to use.

8.4.4 *Determination of Coupling Yield by Alkaline Hydrolysis and Ninhydrin Reaction*

Determine the amount of protein bound on the gel as follows.

(i)   After treatment with chaotropic agent and exhaustive washing with PBS, weigh two samples (300 mg) of gel (interstitial liquid removed) into duplicate test tubes.

(ii)   Add PBS (0.1 ml) and 1.0 M sodium hydroxide (3 ml) to the samples.

(iii)   Incubate at 37°C for 20 h, keeping the gel beads suspended by shaking.

(iv)   Make up a set of controls with unreacted Sepharose, by adding PBS (0.1 ml) containing different amounts of protein (usually BSA, 0, 0.2, 0.4, 0.8, 1.2 and 2.0 mg).

(v)   After hydrolysis add glacial acetic acid (0.22 ml) to each sample and react an aliquot (1 ml) with freshly prepared ninhydrin reagent (1 ml) [modified after (8,9): dissolve ninhydrin (2 g) and hydrindantoin (0.15 g) in methylcellosolve (75 ml); add 4.0 M sodium acetate buffer, pH 5.5 (25 ml)].

(vi)   Dilute all samples with ethanol (2 ml), which clarifies any turbidity, and read the optical density at 570 nm.

8.4.5 *Example A: Sheep Anti-human Fc-Sepharose CL 6B*

Add activated gel (13 g) to a solution of sheep anti-human Fc antibody (10 ml) containing 10.7 mg/ml. React for 48 h at $2-8°C$. The resulting gel contains 4.2 mg of protein per gram of gel (moist weight).

8.4.6 *Example B: Human F(ab')$_2$-Sepharose CL 6B*

Add activated gel (12 g) to pepsin-derived fragment F(ab')$_2$ of human IgG, (10 ml) containing 5.4 mg/ml. React for 48 h at $2-8°C$. The resulting gel contains 4.2 mg of protein per gram of gel (moist weight).

8.4.7 *Example C: Human IgG-Sepharose CL 6B*

(i)   Wash activated gel (57 g) with high ionic strength coupling buffer (0.1 M phosphate containing 1.0 M sodium chloride and 0.01 M carbonate, pH 9.5).

(ii)   Add human IgG solution (50 ml) containing 10 mg/ml (human IgG fraction II, Fluka AG, Buchs, Switzerland) in the same buffer.

(iii)   Allow the reaction to proceed at room temperature for 16 h.

(iv)   Process the gel as above. A gel is obtained which contains 5.5 mg of protein per gram of gel (moist weight).

8.4.8 *Example D: Human Albumin-Sepharose CL 6B*

Add activated gel (19.5 g) to 15 ml of a solution of human albumin (15 mg/ml) in 0.01 M sodium carbonate, pH 9.5 (adjust the pH with NaOH). React for 48 h at $2-8°C$ and process as above. The resulting gel contains 12.5 mg protein per gram of moist gel.

9. POLYISOCYANATE ACTIVATION OF HYDROXYL GROUP-CONTAINING POLYMERS

by G.Krisam

9.1 Introduction to Matrices Used

9.1.1 *Beaded Cellulose*

Beaded cellulose (Institute of Macromolecular Chemistry, Prague, Czechoslovakia) can be produced in the following size ranges: $20-60$, $40-100$, $100-200$, $200-500$ and $500-2000$ μm. High mechanical stability can be achieved without cross-linking of the material due to highly structured regions in the matrix. Porosity is $80-90\%$ with pore sizes of $10-60$ nm. $10-20\%$ of the pores are larger than 60 nm. The production procedure is rather simple and includes the steps shown in *Figure 5*.

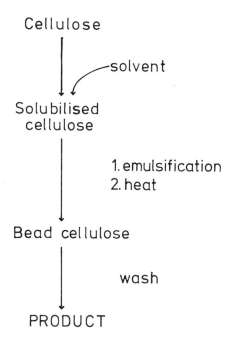

Figure 5. Preparation of beaded cellulose.

9.1.2 *Polyhydroxy-derivatised Controlled Pore Glass (Serva, Heidelberg, FRG)*

Controlled pore glass is composed of spherical particles having diameters of $100-200$ μm. The pore size is $40-80$ nm. Due to the polyhydroxy coating, the material is hydrophilic and electroneutral. Acidic groups in the glass are blocked and it is claimed that non-specific interaction with proteins is negligible.

9.2 **Activation Procedure using Beaded Cellulose as an Example**

(i)   Dehydrate beaded cellulose (preferably $200-500$ μm diameter) with an organic solvent (e.g., acetone or methanol).
(ii)   Add catalyst.
(iii) Remove the solvent and react the beads with polyisocyanate.
(iv)   Wash the beads with further organic solvent, dry *in vacuo* and store in sealed glass bottles.

Activated beaded cellulose is extremely stable. The material can be stored up to 3 weeks in an open glass vessel without appreciable loss in binding capacity. If the material is stored in screw capped glass bottles, no loss in binding capacity has been observed over more than one year. The activation procedure is shown in *Figure 6*.

9.3 **Coupling Procedure**

The coupling of proteins and ligands to the activated matrix is simple (*Figure 7*). Wash the dry pre-activated matrix with 0.1 M HCl, then pour into a solution con-

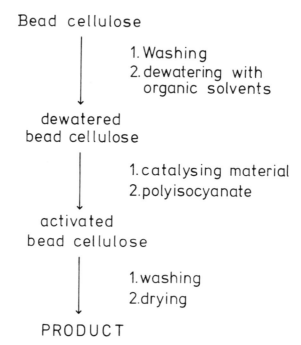

Figure 6. Preparation of polyisocyanate-activated beaded cellulose.

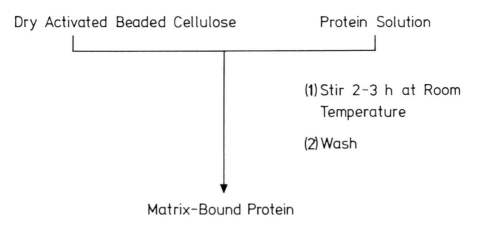

Figure 7. Coupling of albumin to polyisocyanate-activated beaded cellulose.

taining the protein dissolved in an appropriate buffer. The isocyanate groups react with amino groups of the protein. Stir for $2-3$ h at room temperature then wash the product. The reaction is complete within 3 h in the case of albumin (*Figure 8*).

9.4 Comparison with Other Binding Methods

Polymeric spacers and the gentle reaction procedure are suggested to be respon-

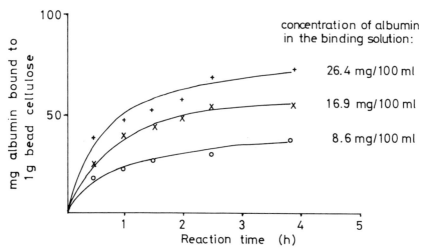

Figure 8. The time-dependence of the reaction of albumin with polyisocyanate-activated beaded cellulose. Polyisocyanate-preactivated beaded cellulose (1 g) was washed with 0.1 M HCl and added to three different concentrations of bovine serum albumin. Aliquots were removed at the times indicated and the concentration of albumin was determined. ($\bullet$——$\bullet$) 26.4 mg albumin per 100 ml, (x——x) 16.9 mg albumin per 100 ml, ($\bigcirc$——$\bigcirc$) 8.6 mg albumin per 100 ml.

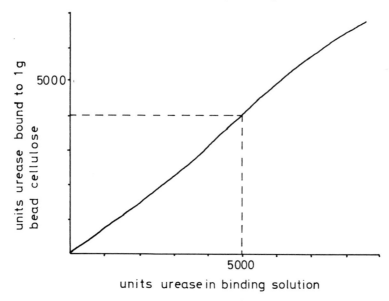

Figure 9. Dependency of the activity of urease-coated beaded cellulose on the activity in the binding solution.

sible for the high yields and low loss of enzyme activity which can be achieved with this method. Urease can be bound in a highly active form; the extent of binding is directly proportional to the concentration of urease in the binding solution (*Figure 9*). The loss of activity of the final product was less than 30% after one year when stored in 50% glycerol at 4°C. A comparison of the polyisocyanate ac-

Table 7. Comparison of Different Matrices for the Immobilisation of Urease and Albumin.

	Urease bound to 1 g	*Albumin bound to 1 g*	*Comments*
Polyisocyanate-activated beaded cellulose	Up to 7200 units	105 mg	No dependence on buffer concentration
Polyisocyanate-activated hydroxyl porous glass	3780 units	40 mg	No dependence on buffer concentration
Polyisocyanate-activated HA-Ultrogel	1000 units	not measured	Activation procedure and binding procedure not optimised for this matrix
Eupergit C	180 units	0	0.1 M phosphate buffer, pH 7.0
	1190 units	16 mg	0.5 M phosphate buffer, pH 7.0
	2400 units	70 mg	1.0 M phosphate buffer, pH 7.0

Table 8. The Activity of Different Enzymes Immobilised to Beaded Cellulose Compared with Commercially Available Polyacrylamide and Agarose-immobilised Enzymes.

Enzyme	*Activity in solution (units)*		*Activity (units per g matrix)*			
	Binding solution	*After binding procedure*	*Cellulose*	*Polyacrylamide (Boehringer-Enzygel)*	*Polyacrylamide (Sigma)*	*Agarose (Sigma)*
Urease	9500	2300	7200	500	50 – 60	n.a.
Glucose oxidase[b]	6000	2000	1100	300	10 – 20	n.a.
Alcohol dehydrogenase[a]	7500	n.d.	7000	n.a.	50 – 100	1000 – 1500
Trypsin[a]	4560	370	1950	30	n.a.	n.a.
β-Galactosidase[a]	460	50	240	n.a.	n.a.	n.a.

[a]Binding procedure not optimised.
[b]Bound in glycine buffer, pH 3.0
1 unit = amount of enzyme which catalyses the reaction of 1 μmol substrate in 1 min at 25°C.
n.a. = not available; n.d. = not detectable.

tivation method has been made with three supports; beaded cellulose, HA-Ultrogel (IBF-Reactifs, Villeneuve-la Garenne, France) and polyhydroxy glass. Urease and albumin were bound to these materials and to Eupergit C (Röhm Pharma GmbH). *Table 7* shows the superiority of the isocyanate activated material with respect to the enzyme activity obtained and the amount of albumin bound to the matrix. Furthermore, no dependency on the buffer concentration was observed with the polyisocyanate-activated materials. A comparison of other commercially available gels with bound enzymes is shown in *Table 8*.

10. IMMOBILISATION USING ORGANIC SULPHONYL CHLORIDES

10.1 **Introduction**

Organic sulphonyl halides react with alkyl hydroxy compounds to form sulphonyl esters which are themselves excellent leaving groups. Nucleophiles will displace the sulphonate by S_N2 type reactions enabling the coupling of molecules of biological interest which are to be immobilised. In principle, any of the hydroxylated supports can be used for this reaction. Both amino- and thiol-containing compounds may be coupled using this procedure. The reactivity of the sulphonate ester is strongly influenced by the substituent attached to the sulphonate. Thus CF_3- > CF_3CH_2- > p-NO_2-phenyl- > CH_3-phenyl- react in the order 4000 : 100 : 20 : 1 (10,11). Tresyl (CF_3CH_2-) and tosyl (p-CH_3-phenyl-) are both convenient reagents for enzyme immobilisation. The former permits efficient nucleophilic displacement at neutral pH and 4°C. Tosyl group leaving can be followed by u.v. absorbance and is cheap; however, coupling must be carried out at pH 9 – 10.5. Water-insoluble ligands can also be easily coupled in organic solvents (e.g., acetone). Tresylated supports can be stored for several weeks after activation in the wet state in 1 mM HCl without harm.

10.2 **Method**

10.2.1 *Activation*

(i)   Wash Sepharose CL 6B (25 g moist weight) by suspending sequentially in water (75 ml), water-acetone (3:1) (75 ml), water-acetone (1:3) (75 ml) and finally acetone (75 ml).

(ii)   With the beads in acetone (75 ml), add tosyl chloride (7 g) dissolved in acetone (10 ml).

(iii)   Stir slowly (overhead type stirrer) and add pyridine (7 ml) drop by drop.

(iv)   Stir (end-over-end type stirrer) at room temperature for 2 h.

(v)   Thoroughly wash the beads with acetone, acetone-water and water using the reverse of the procedure above.

10.2.2 *Coupling*

(i)   To the beads prepared as in Section 10.2.1 add a 1.5 M solution of the amine in 0.5 M $NaHCO_3$, pH 10.7 (final pH) (50 ml).

(ii)   Mix the suspension at 40°C for 16 – 48 h.

(iii)   After coupling has ceased (remove aliquots periodically and follow the release of tosyl groups spectrophotometrically), filter on a sintered glass funnel and wash sequentially with water, 1 mM HCl, 1.0 M $NaHCO_3$ and water (500 ml each).

11. FUNCTIONALISATION OF AGAROSE BEADS VIA CARBOXYMETHYL-ATION AND AMINOETHYLAMIDE FORMATION

by J.K.Inman

11.1 **Introduction**

Primary functionalisation of agarose beads for the preparation of affinity adsorbents has often been carried out by starting with cyanogen bromide activation. The reactive cyanate ester and imidocarbonate groups form mainly isourea linkages with amino groups of proteins or other ligands (12). Under normal conditions of use, these linkages have the disadvantages of being protonated (and thereby charged) and unstable to displacement by nucleophiles (e.g., OH^-). Thus, non-specific ionic interactions can occur and bound ligands gradually break loose from the support. Several alternative approaches to the functionalisation of polysaccharide matrices have been proposed (13 – 16) which do not necessarily suffer these drawbacks.

In this section, one such approach is presented that begins with a practical method for introducing stable, carboxymethyl ether groups into cross-linked agarose beads. Subsequently, the terminal carboxyl functions may be activated for combination with specific ligands or spacing structures, or may be converted to generally useful, primary amino groups by reaction with ethylenediamine as described in Section 11.2.3.

Carboxymethylation of support matrices that possess either primary or secondary hydroxyl groups can be carried out by the classical Williamson ether synthesis as was notably done in the case of CM-cellulose by Peterson and Sober (17). Vicinal diol configurations are not required, and the method should be applicable to all polyhydroxy matrices that can withstand strong alkali treatment for a few hours, such as cellulose and agarose, and cross-linked dextrans, poly(hydroxyethyl) methacrylates or polyvinyl alcohols. The reaction is carried out with chloroacetic acid in an excess of aqueous sodium hydroxide as shown in Equation 1.

$$\text{H}\!-\!\text{O}^- \text{Na}^+ + \text{Cl}\!-\!\text{CH}_2\!\text{C}\!-\!\text{O}^-\,\text{Na}^+ \xrightarrow{\text{NaOH}} \text{H}\!-\!\text{O}\!-\!\text{CH}_2\text{C}\!-\!\text{O}^-\,\text{Na}^+ + \text{Na}^+\text{Cl}^- \qquad (1)$$

Agarose / CM

For most purposes, only a small proportion of the hydroxyl groups needs to be converted. In the procedure below, a large excess of chloroacetate is employed in order to reduce the time of exposure of the matrix to strong alkali and to allow roughly first-order kinetics. This latter circumstance permits easy control of the concentration of introduced carboxymethyl (CM) groups by use of a specified reaction time and temperature (18). The principal side reaction, which does not affect the matrix, is the conversion of chloroacetate to glycolate by attack on hydroxyl ions. In the procedure given here, only about 12% of the reagent is lost by this route before the derivatisation reaction is terminated.

In the next step, a support bearing primary amino groups can be prepared according to Equation 2.

$$\text{H}\!-\!\text{O}\!-\!\text{CH}_2\!-\!\text{C}\!-\!\text{OH} + \text{H}_2\text{N}\!-\!\text{CH}_2\text{CH}_2\!-\!\overset{+}{\text{N}}\text{H}_3\ \text{Cl}^- \xrightarrow{\text{EDAC}} \text{H}\!-\!\text{O}\!-\!\text{CH}_2\text{C}\!-\!\text{N}\!-\!\text{CH}_2\text{CH}_2\!-\!\overset{+}{\text{N}}\text{H}_3\ \text{Cl}^- \qquad (2)$$

CM / AECM

A very large excess of ethylenediamine is employed in order to promote a one-ended reaction: a single amide link is formed leaving an uncombined amino group. A water-soluble carbodiimide (EDAC, see below) is employed as the condensing agent. The reactive species of ethylenediamine is the singly protonated form as shown in Equation 2; however, most molecules are doubly protonated at the controlled reaction pH of 4.6. The nearby positive charge on the other nitrogen atom suppresses the pK_a of the reacting amino group to a value of around 7.1. At pH 4.6 the reactivity of this amino group is thus considerably enhanced with respect to the amino group remaining after attachment (pK_a ~9.2), another factor favouring one-ended linkage to the matrix. Over 90% of the carboxyl functions can be converted to the aminoethylcarbamylmethyl (AECM) derivative.

Together, the reactions of Equations 1 and 2 yield amino groups linked to the matrix by a simple, hydrophilic structure that is completely stable under conditions normally used in affinity separations. The additional derivatisation step described in Section 11.2.4 yields a thiol support, 3-mercaptopropionyl-AECM-agarose, that may also be used in the prepartion of specific adsorbents. Its structure may be represented as $HS-CH_2CH_2-CO-AECM$-agarose. The uncharged, 9-atom spacing structure is relatively hydrophilic, since it possesses only two consecutive CH_2 groups and no aromatic structure.

11.2 **Procedure**

11.2.1 *Materials and Solutions*

The beaded, cross-linked agarose used in the procedures below is Sepharose CL 4B from Pharmacia Fine Chemicals. Chloroacetic acid may be obtained from Sigma Chemical Co. or Fisher Scientific Co. and should be titrated to determine its actual equivalent weight; commercial materials have a small but variable water content.

(i)   Weigh out a sample of chloroacetic acid (~0.8 g) and dissolve in 50 ml water.

(ii)   Add a few drops of bromothymol blue indicator solution [0.1% (w/v) in 50% (v/v) ethanol] and titrate with standard 1.0 or 2.0 M NaOH.

(iii) Calculate the equivalent weight as weight of acid sample in milligrams divided by the milliequivalents of NaOH used.

6.0 M NaOH may be prepared from 10.0 M solution obtained from Fisher Scientific Co.

Ethylenediamine dihydrochloride (EDA.2HCl), 1.5 M (20% w/v), pH 4.6, is made up as follows.

(i)   Dissolve 300 g EDA.2HCl (Sigma or Eastman) in 1.2 litres of distilled water. Make up to 1.5 litres with water.

(ii)   Stir the solution with 15 g of activated charcoal (e.g., Darco G-60) for 30 min.

(iii) Filter through a fluted Whatman No.1 filter paper.

(iv)   Adjust the pH to 4.6 with 5.0 M NaOH.

Unused solution may be kept in an amber glass bottle for many months.

EDAC, 1-ethyl-3-(3-dimethylaminopropyl) carbodiimide hydrochloride, may

55

be purchased from Sigma. SPDP, 3-(2-pyridyldithio)propionic acid N-hydroxy-succinimide ester [also named N-succinimidyl 3-(2-pyridyldithio) propionate], may be obtained from Sigma or Pharmacia.

HE (HEPES-EDTA) buffer has the composition: HEPES (N-2-hydroxyethyl-piperazine-N'-2-ethanesulphonic acid) 0.15 M, NaOH 0.079 M, EDTA (ethylenediamine tetraacetate, di- or trisodium) 1.0 mM; pH 7.6.

SHE (solvent-HEPES-EDTA) buffer is a mixture of 3 volumes of HE buffer and 2 volumes of N,N-dimethylformamide (Spectro grade).

11.2.2 *Carboxymethylated (CM-) Agarose Beads*

Glassware should be siliconised in order to minimise the adherence of agarose beads. The alkaline suspension below is cooled to 15°C before adding chloroacetic acid to compensate for the heat of neutralisation which raises the temperature to 25°C, the desired control point. A suspended, motor-driven stirrer should be employed since magnet bar mixing is apt to fragment the beads. The rapid initial washing serves to terminate the reaction. If 7 min is required to effect transfer to the funnel plus washing out of a major portion of the reagents, then the total reaction time at 25°C is estimated at 75 min (68 + 7). Caution should be exercised in handling chloroacetic acid and the caustic mixture in the procedure below. Protective gloves and glasses should be worn to prevent contact of material with skin or eyes. The steps are as follows.

(i)   Measure into a large beaker a suspension of agarose beads corresponding to about 400 ml of settled bed volume.

(ii)   Suspend beads in 0.1 M NaCl to a volume of 2.4 litres. Stir the suspension for 5 min and allow beads to settle for 60 – 90 min. Siphon off the supernatant down to the 600 ml mark and discard it along with the fines.

(iii)   Repeat step (ii) three more times.

(iv)   Transfer the settled beads to a smaller beaker and adjust the total volume to 600 ml using more 0.1 M NaCl.

(v)   Add 600 ml of 6.0 M NaOH and stir the mixture in an ice bath until its temperature reaches 15°C.

(vi)   Remove from the ice-bath and add 1.20 mol (= equivalents on basis of titrated equivalent weight; see above) of chloroacetic acid.

(vii)   Adjust the temperature to 25.0±0.2°C and maintain stirring at this temperature for 68 min.

(viii)   Immediately transfer the suspension to a No.4A porcelain Buchner funnel, lined with a wet Whatman No.1 filter paper, with suction applied. Wash the gel rapidly with 2.5 litres of 0.1 M NaCl.

(ix)   Wash the gel slowly on the funnel with 10 litres of 0.1 M NaCl employing increasing periods without applied suction (gravity flow).

The resulting CM-agarose may either be stored at 4°C in 0.1 M NaCl containing 0.04% (w/v) sodium azide as preservative or allowed to remain on the funnel if the preparation of AECM-agarose is to be carried out immediately.

11.2.3 *Aminoethylcarbamylmethylated (AECM-) Agarose Beads*

Start with the CM-derivative remaining in the Buchner funnel (~370 ml settled bed volume).

(i)   Wash the gel with four 150 ml portions of 1.5 M ethylenediamine dihydrochloride (Section 11.2.1). Apply suction briefly after each portion has run through.

(ii)   Transfer the suction-drained gel to a 1 litre beaker. Suspend beads in 1.5 M ethylenediamine dihydrochloride to a total volume of 650 ml.

(iii)   Adjust the pH to 4.6 (using 2.0 M HCl or NaOH) and add 5.85 g of EDAC in portions, with stirring, over a period of 30 min at room temperature, and continue stirring for a further 30 min. Keep the pH at 4.6 – 4.8 (2.0 M HCl or NaOH) throughout this step and the next.

(iv)   Add another 5.85 g of EDAC as in the previous step, but continue the stirring and pH control for 4 h after the addition is completed.

(v)   Transfer the suspension to a No.4A Buchner funnel lined with Whatman No.1 paper. Wash the beads with 8 litres of 0.1 M NaCl with increasing periods of gravity flow (no suction). Continue washing with 4 litres of 0.4 M NaCl by gravity flow and finally with 4 litres of 0.1 M NaCl.

(vi)   Store the resulting AECM gel in 0.1 M NaCl containing 0.04% (w/v) sodium azide at 4°C.

The author has kept AECM cross-linked agarose beads for long periods of time (>6 months) at −20°C after exchanging the 0.1 M NaCl for 2-propanol, by washing the gel on a Buchner funnel successively with several bed volumes of 25%, 50%, 75% (v/v) 2-propanol in distilled water, and finally 100% 2-propanol. Aliquots may be removed as required (2-propanol does not freeze at −20°C), warmed to room temperature and washed with a desired buffer.

11.2.4 *Determination of Amino Groups*

A sample of AECM-agarose is treated with an excess of SPDP (19) in a buffer-solvent mixture to convert all amino groups to 2-pyridyldithiopropionamide functions. Subsequent treatment with excess dithiothreitol (DTT) reduces the included disulphide bond and releases pyridine-2-thione into solution as a chromophore; its concentration is measured spectrophotometrically (ϵ_{max} = 8080 at 343 nm) (19).

(i)   Transfer a sample of AECM-agarose gel (3.5 – 4 ml bed volume) to a small Buchner funnel lined with Whatman No.1 paper.

(ii)   Wash with 0.1 M NaCl and then with SHE buffer (see Section 11.2.1). Resuspend in 8 ml of SHE buffer.

(iii)   Dissolve 9 mg SPDP in 0.5 ml of N,N-dimethylformamide and add this solution to the above suspension. Stir at room temperature for 60 min (a magnet bar is satisfactory for an analytical experiment).

(iv)   Wash gel on a Buchner funnel with SHE buffer, 0.1 M NaCl, and finally HE buffer (see Section 11.2.1).

(v)   For a duplicate assay, place 1.0 – 1.5 ml bed volume of (iv) in each of two siliconised 10 x 75 mm test tubes using HE buffer as a medium. Pack the gels by centrifugation for 2 min at approximately 1500 *g*. Allow the tubes to stand for at least 10 min, then mark the position of the bed top for subsequent volume measurement. Remove supernatant buffer.

(vi) Prepare fresh 0.2 M DTT in HE buffer. Add an equal volume of this solution to each gel sample and mix intermittently for 45 min at room temperature.

(vii) Quantitatively transfer the suspensions to 25 ml measuring cylinders and make up to the mark with HE buffer. Mix for several minutes, allowing enough time for the released pyridine-2-thione to diffuse out of the beads.

(viii) Centrifuge samples of (vii) to remove all beads. Measure absorbance (1 cm path) of the clear supernatants at 343 nm using HE buffer as a blank.

(ix) Calculate the concentration (density) of reactive amino groups using the equation μmol/ml of bed $= 25\, A_{343}/(8.08 \times$ ml of gel$)$.

The above procedure for the preparation of AECM-Sepharose CL 4B yields a product with 1.40 μmol of amino groups/ml of gel bed. Higher or lower amino group densities can be obtained by proportionately adjusting the reaction time for carboxymethylation, increasing the time even more than given by this calculation for appreciably higher densities.

11.3 **Concluding Remarks**

The above procedures describe a practical approach to the initial functionalisation of agarose beads for the subsequent preparation of specific affinity adsorbents. The group density of carboxyl, amino or thiol (sulphydryl) groups can best be controlled by setting the time and temperature of the first reaction (Equation 1): the next reaction (Equation 2) and the thiolation reaction of the analytical experiment (Section 11.2.4) are carried out essentially quantitatively. This approach yields useful anchoring functions that are bound in completely stable, uncharged, hydrophilic linkages to the support matrix.

The subsequent attachment of specific ligands can be brought about by a variety of means including activation of the CM derivative with a water-soluble carbodiimide, acylation of the AECM form, or alkylation of AECM amino or 3-mercaptopropionamide groups. The final ligand density may be controlled by the CM density if all subsequent reactions are quantitative, or else, by limiting the final derivatisation. In the latter case, it is usually wise to 'cap' the remaining anchoring groups (e.g., charged, protonated amine). Unused amino groups can be acetylated; SH groups may be alkylated with iodoacetamide at pH 8.0, preferably after a fresh reduction of spontaneously formed disulphide bonds.

12. REFERENCES

1. Longstaff,C. (1983) Ph.D. Thesis, Liverpool University.
2. Cuatrecasas,P. (1980) *J. Biol. Chem.,* **245**, 3059.
3 Porath,J., Aspberg,K., Drevin,H. and Axen,R. (1973) *J.Chromatogr.,* **86**, 53.
4. Lamed,R., Oplatka,A. and Reisler,E. (1976) *Biochim. Biophys. Acta,* **427**, 688.
5. Fischer,E.A. (1982) in *Affinity Chromatography and Related Techniques*, Gribnau,T.C.J., Visser,J. and Nivard,R.J.F. (eds.), Elsevier, Amsterdam, poster A15.
6. Wilson,M.B. and Nakane,P.K. (1976) *J. Immunol. Methods,* **12**, 171.

7. Fischer,E.A. (1983) in Poster 55 presented at the *5th International Symposium on Affinity Chromatography and Biological Recognition*, June 12-17, 1983.
8. Moore,S. and Stein,W.H. (1954) *J. Biol. Chem.,* **211**, 893.
9. Moore,S. and Stein,W.H. (1954) *J. Biol. Chem.,* **211**, 907.
10. Gribnau,T.C.J. (1977) Ph.D. Thesis, Nijmegen, The Netherlands.
11. Nilsson,K. and Mosbach,K. (1981) *Biochem. Biophys. Res. Commun.,* **102**, 449.
12. Kohn,J. and Wilchek,M. (1983) in *Solid Phase Biochemistry*, Scouten, W.H. (ed.), John Wiley & Sons, New York, p.599.
13. Matsumoto,I., Mizuno,Y. and Seno,N. (1979) *J. Biochem.,* **85**, 1091.
14. Bethell,G.S., Ayers,J.S., Hancock,W.S. and Hearn,M.T.W. (1979) *J. Biol. Chem.,* **254**, 2572.
15. Nilsson,K. and Mosbach,K. (1980) *Eur. J. Biochem.,* **112**, 397.
16. Inman,J.K. (1982) in *Affinity Chromatography and Related Techniques*, Gribnau,T.C.J., Visser,J. and Nivard,R.J.F. (eds.), Elsevier Scientific Publishing Company, Amsterdam, p.217.
17. Peterson,E.A. and Sober,H.A. (1956) *J. Am. Chem. Soc.,* **78**, 751.
18. Inman,J.K. (1975) *J. Immunol.,* **114**, 704.
19. Carlsson,J., Drevin,H. and Axen,R. (1978) *Biochem. J.,* **173**, 723.

CHAPTER 3

Cross-Linking Agents for
Coupling Matrices to Spacers

1. INTRODUCTION
Homo- and heterobifunctional reagents have been described in most of the standard texts in affinity chromatography. A list of these compounds is given in *Table 1*.

2. METHODS FOR COUPLING NUCLEOPHILIC LIGANDS TO CARBOXYLATE-CONTAINING MATRICES USING CARBODIIMIDES
Water-soluble carbodiimides may be employed in aqueous media or dicyclohexylcarbodiimide may be used in organic solvents such as dioxane, dimethyl sulphoxide or 80% (v/v) aqueous pyridine. An example is given below.

(i)   Add carboxylate-containing material (e.g., carboxyalkyl glass, carboxyalkyl agarose) (2 – 5 ml) to either 1-cyclohexyl-3-(2-morpholinoethyl) carbodiimide or 1-ethyl-3-(3-dimethylaminopropyl) carbodiimide (40 mg) dissolved in water (10 ml) and adjust to pH 4.0.
(ii) Add the ligand to be coupled (10 – 50 mg of protein or 1 μmol to 10 mmol of amine-containing ligand) and stir the mixture at room temperature for at least 1 h (or longer if the ligand concentration employed is very low).
(iii) Terminate the reaction by filtering the beads and wash them with water (10 vols) and 40% (v/v) aqueous dimethylformamide (1 – 2 litres). More rapid washing can be effected if the matrix is compatible with acetone, 100% dimethylformamide, or with reflux conditions.

Water-soluble carbodiimides have also been used in the preparation of DNA cellulose (see Chapter 5, Section 1). A useful washing procedure for glass beads, particularly when employing carbodiimides which are not water soluble, is to reflux the beads for 5 min with each of the above or to extract overnight in a Soxhlet extractor with an appropriate solvent. This reaction is probably not the most suitable if only limited concentrations of ligand are available.

3. PREPARATION OF HYDROXYSUCCINIMIDE AGAROSE
(i) Wash carboxyagarose (e.g., succinylated aminoalkyl agarose) extensively with anhydrous dioxane on a sintered glass funnel under suction.
(ii) Resuspend the agarose in anhydrous dioxane (3 vols) and add N-hydroxysuccinimide, stirring continuously, to yield a solution with a final concentration of 0.1 M N-hydroxysuccinimide.

Table 1. Bifunctional Compounds of Interest in Affinity Chromatography.

p-Benzoquinone	(1)
Bis-(diazobenzidine)	(3)
3,6-Bis-(mercurimethyl) dioxane	
Bis-oxiranes	(3)
Cyanuric chloride	(1)
p,p'-Difluoro-*m,m'*-dinitrophenylsulphone	(1)
Dicyclohexylcarbodiimide	
Dimethyladipimidate	(3)
Dimethylsuberimidate	(3)
Divinylsulphone	
N,N'-Ethylene-bis-(iodoacetamide)	(3)
Glutaraldehyde	(1)
Hexamethylene bis-(maleimide)	(3)
Hexamethylene diisocyanate	(3)
Periodate	(1)
N,N'-1,3-Phenylene-bis-(maleimide)	(1)
Phenol-2,4-disulphonyl chloride	(3)
Tetra-azotised *o*-dianisidine	(1)
Toluene diisocyanate	(1)
Water soluble carbodiimides	(1)
Woodward's K reagent	

References contain useful comparative data for many of the above reagents. The selection of linking reagent depends to a large extent on the sensitivity (to pH, temperature, solvent etc.) of the ligand to be coupled. The sections in this Chapter describe various specific reagents and recipes for their use. The reader should consult the original papers if in doubt as to their specific application.

(iii)   Add N,N'-dicyclohexylcarbodiimide while stirring (final concentration 0.1 M), and continue stirring at room temperature for 70 min.

(iv)   Collect the beads on a sintered glass funnel under suction, and wash with dioxane (8 vols), followed by methanol (>3 vols) (the latter removes the precipitated dicyclohexylurea).

(v)   Wash the beads with dioxane (3 vols) and either use immediately or dry effectively if they are to be stored.

(vi)   To dry, wash with dioxane (3 vols) which has been dried through a column of neutral alumina.

(vii)   Suspend the beads in alumina-dried dioxane (2 vols) and place in a dark bottle with molecular sieves (type 4A or 5A) and seal.

Less than 10% of the N-hydroxysuccinimide ester will be hydrolysed in 4 months using this procedure.

3.1 Coupling Ligands to Hydroxysuccinimide Ester-agarose

(i)   Collect the beads on a sintered glass funnel under suction, wash quickly with ice-cold water, and add to a solution of the ligand in an appropriate buffer (5 vols buffer per vol of gel). Any pH between 5 and 9 may be employed provided that it is compatible with the ligand. Acetate, citrate, phosphate, borate and bicarbonate buffers are all suitable and do not interfere with the coupling reaction. The reaction time can be from 10 min to

6 h, depending on variables including the pH and the ligand used.

(ii) Terminate the reaction by adding either ethanolamine or glycine to a final concentration of 1.0 M. These amines react with any excess activated ester.

(iii) After a further 2 h at room temperature, wash the beads with the buffer that is required for subsequent use of the matrix.

This procedure can be applied to carboxymethyl cellulose, carboxyacrylamide, and carboxy derivates of Eupergit C or controlled pore glass.

4. PREPARATION OF SUCCINYLATED AMINOALKYL AGAROSE

4.1 Preparation of N-alkylamine Agarose

(i) Wash agarose on a sintered glass funnel with distilled water (10 vols).

(ii) Suspend agarose (100 g) in water (80 ml). Add 1.0 M sodium periodate (20 ml).

(iii) Shake the suspension for 2 h at room temperature, preferably in the dark.

(iv) Wash the oxidised agarose with water.

(v) Add the oxidised agarose to 2.0 M alkyl diamine-HCl (e.g., hexamethylenediamine-HCl), pH 5.0 (100 ml).

(vi) After 6 – 10 h at room temperature raise the pH to 9.0.

(vii) Cool the beads to 0 – 4°C and add 5.0 M sodium borohydride (10 ml) over 12 h.

(viii) Wash the beads thoroughly with 1.0 M NaCl by allowing the solution to percolate over the beads on a sintered glass funnel for 3 – 5 h without applying suction.

(ix) Leave the beads to stand in 1.0 M NaCl overnight.

(x) Percolate the beads for 3 – 5 h with 1.0 M NaCl, as in (viii).

4.2 Succinylation

(i) Suspend the aminoalkyl agarose in water (1 vol), and add succinic anhydride (10-fold excess over the amino content of the agarose), maintaining the pH at pH 6.0 with 5.0 M NaOH.

(ii) Monitor the pH constantly until it remains stable for at least 30 min without further addition of NaOH.

(iii) Maintain the temperature between 25°C and 30°C, using an ice-bath if necessary.

(iv) Incubate the mixture for 5 h at 4°C, then filter under suction and with distilled water.

(v) Test a sample of the beads with ninhydrin to determine whether the reaction is complete. If a light blue-purple colour develops, repeat the succinylation procedure using a larger excess of succinic anhydride and leave the reacted suspension for 5 h at room temperature.

5. ACYL AZIDE ACTIVATION

5.1 Preparation of Hydrazido-agarose

(i) Prepare adipoyl hydrazide by heating under reflux for 3 h, diethyladipate

(100 ml) with hydrazine hydrate (200 ml) and ethanol (200 ml). On cooling the reaction mixture to room temperature, adipoyl hydrazine precipitates.
(ii) Recrystallise from ethanol-water (m.p. 169 – 171°C).
(iii) Mix CNBr-activated agarose beads, prepared as described in Chapter 2, with a cold saturated solution of adipoyl dihydrazide in 0.1 M Na₂CO₃ (pH 9.0) (2 vols).
(iv) Shake the beads gently overnight and then wash thoroughly on a sintered glass funnel with distilled water, 2.0 M NaCl and finally distilled water.

5.2 Preparation of Acyl Azido-agarose

Convert the hydrazido-agarose into acyl azide-agarose as follows.

(i)   Suspend the hydrazido-agarose beads in ice-cold 1.0 M HCl (1 vol). To this add ice-cold 0.5 M NaNO₂ (1 vol).
(ii) Maintain the temperature at 0°C for 15 – 20 min, then filter the beads on a sintered glass funnel under suction and wash with ice-cold 0.1 M HCl.
(iii) Empty the filtrate from the suction flask and rapidly wash the beads with ice-cold 0.1 M sulphamic acid and then ice-cold water.
(iv) Suspend the washed beads in 0.2 M NaHCO₃, pH 9.5 (10 vols), containing the ligand that is to be bound. The concentration of the ligand is dependent upon its cost, chemical characteristics and availability.

Proteins are best coupled at concentrations of 5 – 20 mg/ml, whereas smaller amino-ligands may be coupled at concentrations of 10 – 100 mM. For a polyamino compound such as lysine, coupling should be effected through the α- rather than the ε-amino group by lowering the pH of the coupling solution; use 0.5 M phosphate buffer, pH 6.0. Prepare and use acyl azide derivatives of polyacrylamide in essentially the same fashion.

6. IMMOBILISATION OF PROTEINS ONTO POLYACRYLAMIDE BEADS USING GLUTARALDEHYDE

by S.Avrameas

6.1 Preparation and Activation of the Beads

Allow polyacrylamide beads to hydrate for 24 h in distilled water and then wash several times with distilled water. Continue to wash polyacrylamide agarose beads until the supernatant is completely clear.

6.2 Activation of the Beads

(i)   Add sedimented beads (100 ml) to distilled water (250 ml), add 1.0 M phosphate buffer, pH 7.4 (50 ml), and 25% (v/v) glutaraldehyde (100 ml).
(ii) Incubate overnight at 37°C.
(iii) Wash the beads thoroughly with distilled water by successive decantations, using a sintered glass filter, or use successive centrifugations at 4000 r.p.m. (e.g., for Bio-Gel P 300 – 400 mesh), until no odour of glutaraldehyde is detected and the A_{280} of the washing solution is zero.

At this stage, the activated beads can be stored in distilled water at 4°C for at least a month without any appreciable loss of protein-binding capacity.

6.2.1 *Coupling Proteins to Activated Beads*

Add centrifuged beads (10 ml) to the protein to be coupled (10 – 50 mg) in phosphate-buffered saline (PBS) (10 ml), 0.01 M phosphate, pH 7.4, containing 0.15 M NaCl. Agitate the suspension on a rotary mixer overnight at room temperature (or at 4°C if necessary).

6.2.2 *Capacity of Gels*

The maximum amount of protein which can be bound per ml of gel is approximately 3 mg. When a high protein content of the gel is required, incubation of the beads must be performed with 30 – 50 mg protein. On the other hand, when as much protein as possible is to be immobilised, a relatively low concentration of protein should be used (10 – 20 mg).

6.2.3 *Choice of Gel*

Bio-Gel P 300 (50 – 100 mesh) and acrylamide-agarose beads are generally employed for column procedures. Bio-Gel P 300 (400 mesh) beads are recommended for batch procedures. Very similar procedures are described for Magnogel 44 (see Chapter 1, Sections 2.6 and 8 and ref. (4) of this chapter).

7. DIVINYLSULPHONE-BASED COUPLING PROCEDURES

7.1 **Activation of Agarose Beads**

(i)   Wash agarose (10 g moist weight) and remove water under suction.
(ii)   Suspend the agarose in 1.0 M sodium bicarbonate, pH 11.0 (10 ml).
(iii)   Add divinylsulphone (2 ml) to the solution.
(iv)   Continue the reaction with stirring for approximately 1 h at room temperature.
(v)   Wash the beads with water.

BEWARE: divinylsulphone is toxic

7.2 **Coupling of Divinylsulphone-agarose**

(i)   Dissolve the amino- or hydroxyl-containing ligand in 1.0 M sodium carbonate to give a final concentration of 250 mg/ml.
(ii)   Add this solution to activated divinylsulphone agarose (10 g) and stir gently. The time needed for complete reaction depends on the temperature employed (e.g. 2 h at 50°C or 24 h at 4°C).
(iii)   Wash the reacted beads with 1.0 M NaCO$_3$, pH 11.0 (10 ml), then with 0.2 M glycine-HCl, pH 3.0, containing 1.0 M NaCl (500 ml), 1.0 M NaCl (500 ml) and water (500 ml).

8. DIAZONIUM-BASED COUPLING PROCEDURES

8.1 **Preparation of NAD-agarose**

(i) Prepare ice-cold, washed CNBr-activated Sepharose (50 ml) as described in Chapter 2, Section 1.

(ii) Remove the water by suction on a sintered glass funnel.

(iii) Stir overnight at 4°C in 1.0 M hexane-1,6-diamine (100 ml) previously adjusted to pH 10.0 with HCl.

(iv) Thoroughly wash the aminohexyl agarose with distilled water, resuspend in 50% (v/v) aqueous redistilled dimethylformamide and stir for 3 – 4 h at room temperature with 250 ml of a solution of 0.07 M *p*-nitrobenzyl azide in 40% (v/v) dimethylformamide-60% 0.2 M borate, pH 9.3.

(v) Wash the *p*-nitrobenzamidohexyl agarose with 50% (v/v) dimethylformamide (1 litre) and distilled water.

(vi) Stir at 37°C for 1 h in 0.5 M NaHCO$_3$ containing 0.1 M sodium dithionite, pH 8.5 (150 ml).

(vii) Wash the resulting *p*-aminobenzamidohexyl agarose with distilled water (1.5 litres), cool to 0°C and diazotise by stirring for 10 min with 2.5 vol of chilled 0.1 M NaNO$_2$ in 0.5 M HCl.

(viii) Wash the *p*-diazo derivative with chilled distilled water (250 ml), 1% (w/v) sulphonic acid (50 ml) and distilled water (250 ml) and then allow to react overnight at 4°C with 0.1 M NAD in 0.2 M borate, pH 8.3.

(ix) Collect the immobilised NAD derivative of benzamidohexyl Sepharose by filtration and wash thoroughly.

8.2 **Preparation of Diazonium Agarose**

(i) Prepare ice-cold, washed CNBr-activated Sepharose (50 ml) as described in Chapter 2, Section 1.

(ii) Remove the water by suction on a sintered glass funnel.

(iii) Stir overnight at 4°C in 1.0 M benzamidine (100 ml) previously adjusted to pH 10.0 with HCl.

(iv) Thoroughly wash the benzamidine agarose with distilled water and diazotise by stirring for 10 min with 2.5 vol chilled 0.1 M NaNO$_2$ in 0.5 M HCl.

(v) Wash the *p*-diazo derivative with chilled distilled water (250 ml), 1% (w/v) sulphamic acid (50 ml) and distilled water (250 ml) and then allow to react overnight at 4°C with the appropriate ligand in 0.2 M borate, pH 8.3.

(vi) Collect the agarose by filtration and wash thoroughly.

WARNING: benzamidine is highly toxic.

A similar procedure can be used with *m*-aminonitrobenzene. In this case the ligand is immobilised via the amine and the nitro group reduced with dithionite and then diazotised as described above.

9. COUPLING LIGANDS VIA REACTIVE AZO DYES

by T.C.J.Gribnau

9.1 Preparation of Regenerated Cellulose

(i)   Dissolve cotton wool (3.3 g), under nitrogen, in 1 litre of a cupric tetrammonium-hydroxide solution (containing 12 g of cupric hydroxide and 2 g of cuprous chloride in 1 litre of 25% saturated aqueous ammonia).

(ii) Filter the dark-blue viscous solution through glass-wool under nitrogen pressure and then inject, stirring continuously, into 5% (w/v) aqueous potassium sodium tartrate (1 litre); maintain neutrality by continuous addition of acetic acid.

(iii) Collect the precipitated cellulose and wash thoroughly with 5% (v/v) acetic acid and water. The material can be stored in 0.1% (w/v) aqueous NaN_3 at about 4°C.

9.2 Determination of the Cellulose Content of Regenerated Cellulose Suspension

(i)   Dilute the product from Section 9.1 with 0.1% (w/v) aqueous NaN_3 in a beaker up to a cellulose content of about 20 mg/ml. Stir thoroughly until the suspension is homogeneous and immediately withdraw 1 ml into a pre-weighed conical centrifuge tube (e.g., Eppendorf Reaktionsgefäss, 1.5 ml, Type 3810).

(ii) Centrifuge at ~12 000 g (~15 000 r.p.m.) for 3 min; pour the supernatant off (or aspirate carefully) and add distilled water (1 ml).

(iii) Resuspend the cellulose, and centrifuge again.

(iv) Repeat this step three times using distilled water, and afterwards three times with acetone (reagent grade quality).

(v)   Evaporate the excess acetone in the pellet by warming the centrifuge tube ($\leq 56°C$) for about 15 min.

(vi) Dry the tube and its pellet subsequently *in vacuo* over concentrated sulphuric acid, at room temperature overnight.

(vii) Allow the tube and contents to re-equilibrate at room temperature and humidity for 2 h before final weighing.

The cellulose content of the suspension described in this section, determined in this way, should be about 20 mg/ml. Store the suspension at 4°C.

9.3 Coupling of the Reactive Azo-dye to Regenerated Cellulose, and Reduction to Aminoaryl-cellulose

(i)   Stir the cellullose suspension (from Section 9.2) thoroughly, and take out a volume containing about 1 g of cellulose (~50 ml).

(ii) Isolate the cellulose by filtration (sintered glass funnel; e.g., Schott & Gen, Mainz, 3D-3) or centrifugation (~1000 g), and wash thoroughly with distilled water (3 x 50 ml).

(iii) Resuspend the cellulose in distilled water (5 – 10 ml) and adjust the temperature to 40 – 45°C.

(iv)   Add Levafix Brilliant Red E-4Ba (Bayer; 25 mg), dissolved in distilled water (10 ml) to the suspension.

(v)   Stir the suspension during 5 min (40 – 45°C), add NaCl (2.5 g) dissolved in distilled water (10 ml), and stir the suspension for 30 min (40 – 45°C).

(vi)   Add solid, anhydrous, Na_2CO_3 (0.5 g), and stir the suspension during 90 min (40 – 45°C).

(vii) Wash the coloured cellulose free from uncoupled dye with distilled water, on a sintered glass funnel or by centrifugation (~ 1000 g).

(viii) Resuspend the coloured cellulose in distilled water (10 ml), and adjust the temperature to 65 – 70°C.

(ix)   Add solid sodium dithionite ($Na_2S_2O_4$; 0.5 – 1 g) which results in almost immediate discolouration of the cellulose (carry out in a well-ventilated hood); the supernatant remains yellow.

(x)   Isolate the aminoaryl-cellulose by filtration or centrifugation, and wash thoroughly with distilled water.

(xi)   Resuspend the aminoaryl-cellulose in 0.1% (w/v) aqueous NaN_3, and store at 4°C, in the dark.

9.4 Diazotisation of Aminoaryl-cellulose, and Coupling with Proteins

CAUTION: CARRY OUT IN A WELL-VENTILATED FUME HOOD

(i)   Stir the cellulose suspension (from Section 9.3) thoroughly, and take out a volume containing about 1 g of cellulose.

(ii)   Transfer the suspension into a sintered glass funnel (e.g., Schott & Gen, Mainz; 3D-3), remove water by suction, and wash the cellulose thoroughly with distilled water (3 x 40 ml).

(iii) Dry the resulting cellulose cake thoroughly under suction and suspend in 0.1 M HCl (30 ml) which has been pre-cooled to 0 – 4°C.

(iv)   Stir the suspension thoroughly, keeping the temperature at 0 – 4°C.

(v)   To the suspension, add pre-cooled (0 – 4°C) 1.0 M $NaNO_2$ (3 ml); maintain thorough stirring and keep the temperature at 0 – 4°C, for 45 min.

(vi)   Transfer the reaction mixture to a sintered glass funnel, and wash the diazotised aminoarylcellulose thoroughly with pre-cooled (0 – 4°C) distilled water until the pH of the filtrate equals that of distilled water (6.0 – 7.0).

(vii) Wash the cellulose with pre-cooled (0 – 4°C) borate buffer (2 x 40 ml).

(viii) The borate buffer is prepared as follows: dissolve H_3BO_3 (12.36 g) and NaCl (2.92 g) in distilled water (900 ml), add 1.0 M NaOH solution (55.2 ml) and adjust the pH to 8.6 with 1.0 M NaOH or HCl before making up to 1 litre.

(ix)   Filter the cellulose and suspend in the relevant protein solution (NB: the coupling solutions must be free of sodium azide). The protein should be dissolved in the same borate buffer described here; in the case of sheep anti-(rabbit IgG) use about 100 mg of IgG/g cellulose.

(x)   Keep the volume of the final reaction mixture as small as possible in order to obtain maximum coupling efficiency; homogeneous mixing (magnetic stirrer or roller-bank) must, however, remain possible.

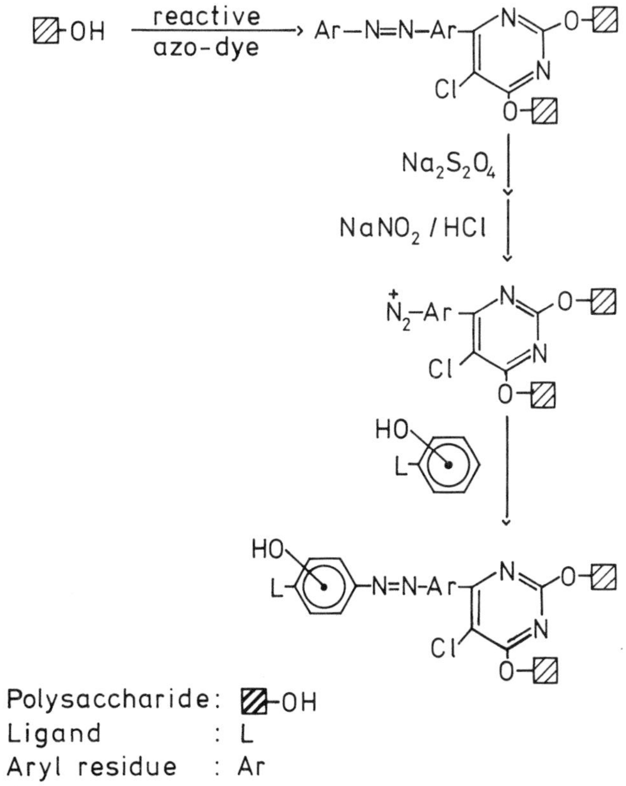

Figure 1. Coupling of HCG to Sepharose CL 4B via reactive azo-dye activation.

(xi)   After 5 – 10 min add 30% (w/v) NaCl (1 ml) to the reaction mixture, and allow the coupling reaction to proceed for 20 – 22 h at 4°C.

(xii)   Filter the cellulose derivative on a sintered glass funnel, and wash twice with distilled water (2 x 40 ml).

(xiii)   Determine the volume of the three filtrates, collected separately, and their protein content; this permits quite an accurate calculation of the degree of protein coupling.

(xiv)   Wash the filter cake with two successive 40 ml portions of a β-naphthol solution (at room temperature); the excess diazonium groups are blocked in this way, yielding an orange-coloured cellulose.

(xv)   The β-naphthol solution should be freshly prepared: add 0.75 g of β-naphthol to 1 litre of borate buffer (as above); stir the suspension for 1 h at room temperature in the dark, and filter the solution subsequently.

(xvi)   Wash the filter cake with two 40 ml portions of borate buffer followed by three 40 ml portions of phosphate buffer; resuspend the cellulose derivative in the same buffer, and store at 4°C.

(xvii) The phosphate buffer is prepared as follows: dissolve in 1 litre of distilled water, sodium merthiolate (1 g) or sodium azide (1 g), $Na_2HPO_4.2H_2O$ (2.46 g), KH_2PO_4 (0.72 g), NaCl (1.125 g) and $Na_2EDTA.2H_2O$ (2.0 g).

(xviii) Determine the concentration of the final suspension by the procedure described in Section 9.2.

9.5 Coupling of HCG to Sepharose CL 4B via Reactive Azo-dye Activation

Coupling is carried out according to the procedure shown in *Figure 1*.

9.5.1 *Coupling of the Reactive Dye to Sepharose CL 4B*

(i)   Add Levafix Brilliant Red E-4BA (Bayer; 125 mg), dissolved in distilled water (12.5 ml) to Sepharose CL 4B (25 g filtered gel).

(ii)   Adjust the temperature of the stirred reaction mixture to 40°C.

(iii)   Add solid NaCl (2.29 g) and stir the suspension for 30 min at 40°C.

(iv)   Add solid, anhydrous, Na_2CO_3 (1.4 g) and stir the suspension for 90 min at 40°C.

(v)   Filter off the Sepharose-dye derivative on a sintered glass funnel and wash thoroughly with distilled water; read the A_{512}^{1cm} of the combined filtrates.

(vi)   The coupling percentage of the dye can be calculated using the value $A_{512}^{1cm,1\%}$, (Levafix Brilliant Red E-4BA) = 176.

(vii) Suspend the coloured Sepharose in 0.2 M aqueous ethanolamine (30 ml), previously adjusted to pH 10.0 with concentrated HCl, and stir the suspension for 30 min at 40°C.

(viii) Filter off the Sepharose and wash thoroughly with distilled water until the pH of the filtrate has reached a value of 6 – 7.

9.5.2 *Reduction of the Sepharose-dye Derivative to Aminoaryl-Sepharose*

(i)   Suspend the final product of Section 9.4.1 in distilled water (30 ml) and adjust the temperature of the suspension to 60°C.

(ii)   Add solid sodium dithionite ($Na_2S_2O_4$; 1.25 g) which results in immediate discolouration of the Sepharose; the supernatant remains yellow.

(iii)   Filter off the Sepharose and wash thoroughly with distilled water (8 x 50 ml).

9.5.3 *Diazotisation of Aminoaryl-Sepharose; Coupling with HCG*

(i)   Suspend the final product from Section 9.4.2 in 0.1 M HCl (45 ml) and add 1.0 M $NaNO_2$ (9 ml).

(ii)   Stir the reaction mixture for 30 min at 4°C.

(iii)   Filter off the diazotised aminoaryl-Sepharose, wash thoroughly with distilled water (4°C) until the pH of the filtrate is 6 – 7, and then wash with borate buffer, pH 8.6 (4°C; 3 x 50 ml).

(iv)   Suspend the diazotised support material (20 g moist weight) in borate buffer, pH 8.6 (4°C; 15 ml); add HCG solution in borate buffer, pH 8.6 (10 ml, 1 mg/ml), and allow the coupling reaction to proceed for 16 h at 4°C (agitation using a roller-bank is advised).

(v)   Filter off the product, wash with borate buffer, pH 8.6 (4 x 50 ml; A) with a
        β-naphthol solution in borate buffer, pH 8.6 (0.75 g/l; 3 x 50 ml) with
        borate buffer, pH 8.6 (5 x 50 ml) and finally with borate buffer, pH 7.0
        containing NaN$_3$ (1 g/litre).

Borate buffer: pH 7.0: Na$_2$B$_4$O$_7$.10H$_2$O   1.43 g/l
                         H$_3$BO$_3$ 11.47 g/l
                pH 8.6: Na$_2$B$_4$O$_7$.10H$_2$O   10.49 g/l
                         H$_3$BO$_3$ 5.57 g/l
                         NaCl   2.93 g/l

β-Naphthol solution (prepare fresh): add β-naphthol (0.75 g) to borate
buffer, pH 8.6 (1 litre), stir the solution for 1 h at room temperature in the
dark, and filter.

(vi)   Store the HCG-Sepharose at 4°C, suspended in this buffer.

Sodium azide should be absent from coupling buffers, because N$_3^-$ groups
substitute the -N$_2^+$Cl$^-$ groups, thereby preventing protein (ligand) coupling. The
HCG content of the combined filtrates A is determined by HCG EIA, yielding
the HCG content of the affinity support; treatment of the Sepharose derivative
with β-naphthol yields an immediate red-orange colouration.

10. SYNTHESIS OF THIOL AGAROSES

10.1 Activated Glutathione-agarose

(i)   Activate 50 g agarose by cyanogen bromide as described in Chapter 2, Sec-
        tion 1 or by one of the alternative methods described in the same chapter.
(ii)   Wash the activated agarose with 0.1 M NaHCO$_3$, pH 8.5 (10 vols) contain-
        ing 20 mg glutathione/g agarose.
(iii)   Shake the mixture for 20 h at 22°C.
(iv)   Wash the gel with 0.1 M NaHCO$_3$, pH 9.5 (10 vol), on a sintered glass fun-
        nel. Transfer to a column (1.8 cm x 30 cm, ~50 ml) and wash sequentially
        with 240 ml of:
        (a)   0.1 M NaHCO$_3$, pH 9.5, containing 1.0 M NaCl and 1 mM EDTA;
        (b)   0.1 M sodium acetate, pH 4.3 containing NaCl and EDTA as before,
              and
        (c)   1 mM EDTA.
        A flow-rate of 10 ml/h for each wash solution is recommended.
(v)   Transfer the gel (50 g, moist weight) to a sintered glass funnel and wash the
        gel with 0.1 M Tris-HCl, pH 8.0 (450 ml) containing 0.3 mM NaCl and
        1 mM EDTA.
(vi)   Remove surplus liquid and suspend the washed gel in the same buffer
        (60 ml) containing 30 mM dithiothreitol (DTT) and incubate for 10 min to
        reduce all available sulphydryl residues.
(vii)   Wash the gel with 150 ml each of 1 M NaCl, 0.5 M NaHCO$_3$, 0.3 M NaCl,
        0.1 M Tris-HCl (pH 9.0), and 0.3 M NaCl, each buffer containing 1 mM
        EDTA.

(viii) Activate the reduced and washed gel by incubating it for 30 min in 1.5 mM 2,2'-dipyridyldisulphide at room temperature, shaking gently.

(ix) Wash the activated glutathione-2-pyridyl disulphide-agarose to free it from excess 2,2'-dipyridyldisulphide by using the same sequence of wash buffers as before.

(x) The final wash with 0.1 M Tris-HCl and 0.3 M NaCl, pH 8.0, containing 1 mM EDTA should be monitored at 280 nm.

(xi) Continue the washing until no further absorbance at 280 nm is detectable in the wash solution.

10.2 Synthesis of Thiopropyl-agarose

(i) Wash 6% agarose (30 g) beads with water and collect on a sintered glass funnel under suction.

(ii) Suspend the gel in 1.0 M NaOH (24 ml). Add epichlorohydrin (0.75 ml) to the stirred suspension at room temperature. [This should yield a final thiol content of 50 μmol/g of agarose (dry weight)].

(iii) Mix at room temperature for 15 min, then raise the temperature to 60°C and continue stirring for 2 h.

(iv) Wash the resulting epoxy-beads with water until neutral, followed by several volumes of 0.5 M sodium phosphate, pH 6.2.

(v) Immediately filter the beads under suction and add to 2.0 M sodium thiosulphate (30 ml).

(vi) Stir the beads in the thiosulphate for 6 h at room temperature, wash with DTT (8 mg/ml) in 1 mM EDTA in at least a 2-fold excess over the epoxide concentration (4 ml of the DTT solution should be sufficient).

(vii) After 30 min of reaction at room temperature, remove the beads from the reduction solution under suction and wash successively with 300 ml each of 0.1 M NaHCO$_3$ containing 1.0 M NaCl and 1 mM EDTA, then 1 mM EDTA, and finally with 0.01 M sodium acetate, pH 4.0, containing 1 mM EDTA. It is important to prevent air from passing over the beads during washing and to protect the beads from further air oxidation by storing them under degassed 0.01 M sodium acetate, pH 4.0. Since an 80% reduction of the concentration of free thiol groups occurs in 1 month (5), reduction of the beads should be performed immediately prior to use. Beads that have been stored for a long period of time can be regenerated to their original capacity by reduction with DTT.

10.2.1 *Activation of Reduced Thiopropyl-agarose*

(i) Wash the beads (which should be freshly reduced to assure a maximal sulphydryl content) with water and then 50% (v/v) acetone-water.

(ii) Suspend the gel in 50% (v/v) acetone-water containing 2-pyridine disulphide (100 mg) dissolved in a minimum of acetone-water.

(iii) Stir the mixture for 30 – 60 min at room temperature and then wash with 50% (v/v) acetone-water.

(iv)   Finally, wash the beads with 1 m M EDTA, pH 7.0. At this point the beads are stable and can be stored for at least 6 months at 40°C with little loss of active thiol content.

11. REFERENCES

1.   Avrameas,S., Ternynck,T. and Guesdon,J. (1978) *Scand. J. Immunol.*, **8**, 7.
2.   Wold,F. (1967) in *Methods in Enzymology*, Vol. **11**, Hirs,C.H.W. (ed.), Academic Press Inc, New York and London, p. 617.
3.   Lowe,C.R. and Dean,P.D.G. (1974) in *Affinity Chromatography*, published by J.Wiley & Sons, London. p. 234.
4.   Hocking,J.D., and Harris,J.I. (1973) *FEBS Lett.*, **34**, 280.
5.   Axen,R., Drevin,H. and Carlsson,J. (1975) *Acta Chem. Scand.*, **29**, 471.

CHAPTER 4

Operational Methodologies

1. DETERMINATION OF LIGAND CONCENTRATION

1.1 Difference Analysis

The quantity of a ligand coupled to a known weight or volume of gel is frequently measured by using the difference between the total amount of ligand added to the coupling mixture and that recovered after extensive washing. In many cases this approach is sufficiently accurate and therefore satisfactory provided that the ligand may be sensitively assayed by a spectrophotometric, fluorimetric or radiometric method. Of course these methods are not the only ones that can be used: n.m.r, e.s.r. and gravimetric measurements are also applicable.

1.2 Direct Spectroscopy

When ligands absorb at wavelengths above 250 nm, it is possible to estimate the concentration of bound ligand by direct spectroscopy of the gel itself. Suspend the gel in optically clear ethylene glycol, glycerol, concentrated sucrose solution or 1% (v/v) aqueous polyethylene glycol (Polyox WSR 301) and read against a similar concentration of control, underivatised gel in a double beam spectrophotometer. It is important that the gels are thoroughly washed with the suspension medium beforehand.

1.3 Solubilisation of Gels

A number of methods are available for solubilising derivatised agarose gels which allow quantitative spectrophotometric estimation of the immobilised ligand. Derivatised agarose gels may be solubilised by warming at 75°C with (i) 0.1 M HCl, (ii) 0.1 M NaOH - 0.1% (w/v) NaBH$_4$ or (iii) 50% (v/v) acetic acid. The exact conditions to achieve optimal solubilisation vary from one preparation to another and are best found by trial and error.

1.4 Acid or Enzymic Hydrolysis

More vigorous treatment of immobilised ligands will hydrolyse the matrix-ligand bond and liberate either free ligand or a degradation product derived from it which may be assayed. Typically, complete hydrolysis of agarose immobilised ligands may be achieved by heating for 1 h at 100°C in 0.5 M HCl. Alternatively, hydrolysis *in vacuo* in 6.0 M HCl for 24 h at 110°C is sufficient to yield products suitable for amino acid analysis and is particularly useful where the immobilised ligand is an amino acid or protein.

1.5 Elemental Analysis

In those cases where the immobilised ligand contains an unique element, elemental analysis can give unambiguous estimates of the ligand concentration. Thus, phosphate analysis is particularly useful for immobilised nucleotides and nucleic acids and sulphur analysis has been used to estimate sulphanilamide coupled to CNBr-activated agarose (1). Elemental nitrogen, bromine or phosphorus analyses can be used despite the fact that steps such as cyanogen bromide activation contain some or all of these elements. However, note: immobilised phosphate can be introduced into agarose gels when CNBr activations are carried out in the presence of phosphate buffer.

Preliminary results suggest that this method may work for Eupergit when immobilising ligands such as butylamine.

1.6 Radioactive Methods

One of the most sensitive methods for determining the immobilised ligand concentration is to incorporate a radiolabelled ligand in the coupling step. The immobilised ligand concentration may be determined by difference analysis, by hydrolysis or by direct measurements on the gel. For example, resuspend aliquots of a gel containing [^{32}P]GTP in Bray's solution and count directly in a liquid scintillation counter (2). Another example of a useful application of this technique is that of ^{131}I-labelled proteins which may be counted directly.

1.7 2,4,6-Trinitrobenzenesulphonic Acid (TNBS) Method for the Determination of Amines

1.7.1 *Materials*

Crystalline TNBS (Eastman), and lyophilised preparations of bovine serum albumin (Sigma) are used without further purification. All amino acids, amines and salts are reagent grade. Dissolve amino compounds in 0.1 M sodium tetraborate, pH 9.3.

1.7.2 *Procedure*

(i)   Add 0.03 M TNBS (25 μl) to the sample (1 ml) contained in a cuvette, agitate to ensure complete mixing and allow to stand for 30 min at room temperature (25°C).

(ii)   Prepare a reagent blank consisting of 0.03 M TNBS in 0.1 M borate (1 ml).

(iii) Read absorbance at 420 nm.

1.8 Determination of Alkylamino Groups Coupled to CNBr-activated Agarose

p-Nitrophenolate ester is commonly used to monitor the nucleophilicity of various compounds. Thus, 1,6-diaminohexane causes the release of two equivalents of *p*-nitrophenolate from *p*-nitrophenolate ester. At concentrations of 1,6-diaminohexane smaller than that of the ester, the rate of liberation of *p*-nitrophenolate from the ester obeys pseudo-first-order kinetics with respect to the diamine up to an ester:diamine ratio of 2.

After all the diamine has reacted, it is possible to estimate its concentration by extrapolating the rate of spontaneous hydrolysis to zero time. This technique for the determination of amine concentration can also be applied to a Sepharose-bound diamine.

1.8.1 *Procedure*

(i)   Wash a sample of coupled gel (50 – 200 mg) with 0.25 M borate buffer, pH 8.5, and transfer to a 1 ml Eppendorf microtube, containing a small magnetic stirring bar.

(ii) Make up the volume to 0.22 ml with the same buffer and add 0.1 M *p*-nitrophenyl acetate in acetonitrile (0.2 ml) at time zero.

(iii) Stir the suspension at room temperature and remove aliquots of 10 μl approximately every 10 min over 2 h.

(iv) Dilute the aliquots in ice-cold 0.05 M borate buffer, pH 8.75 (5 ml) and read the absorbance rapidly at 400 nm.

(v)   To the residual suspension add 2.5 M NaOH (5 μl) and sample the final aliquot and treat as above.

The absorbance of the final aliquot represents the total potential absorbance of the sample. Use the buffer and untreated agarose as blanks.

2. LIGAND LEAKAGE

A serious limitation in the application of affinity chromatography is in systems with a high affinity (for example, hormone-receptor interactions).

The isolation of picomole and nanomole amounts of proteins is compromised by the relatively easy release of affinity ligands into the solution. These problems arose historically primarily because affinity ligands were bound monovalently through a spacer on to cyanogen bromide-activated agarose, in the presence of compounds (e.g., buffers and proteins) that contain nucleophiles.

In order to elucidate the leakage of affinity ligands from specific adsorbents more closely, Tesser *et al.* (3) prepared adsorbents by binding adenosine 3′,5′-cyclic monophosphate and other substances through various spacers to agarose, cellulose and cross-linked dextran by means of cyanogen bromide activation, and to polyacrylamide by an amide bond formed with the carboxyl groups of the carrier, according to the method described by Inman and Dintzis (4). From the dependence of the amount of detached ligands on pH and time, it follows that the leakage of affinants from the matrices takes place predominantly at the site of the fixation on the surface of the solid carrier, and that it can be enhanced with the assistance of neighbouring carboxyl and carboxamide groups in polyacrylamide gels and hydroxyl groups in agarose, cellulose and cross-linked dextrans. The releasing reaction is general and independent of the structure of the bound affinity ligand. The ligands bound to polyacrylamide gels by the R-NH-CO-acrylamide gel bonds are detached more slowly.

The leakage of bound ligands can be suppressed if binding through polyvalent spacers is used. A further improvement can be achieved by bonding ligands to the carrier more tightly, such as by the periodate oxidation method or via epoxide-activated agarose. The variation in the degree of leakage has been studied as a function of the carrier used. In the isolation of oestrogen and insulin receptors, for example, it is impossible to use glass and polyacrylamide carriers, owing to the rapid leakage of affinity ligands.

3. LIGAND CONCENTRATION EFFECTS

The earliest study of the effects of ligand concentration on the binding of enzymes to affinity ligands was by Harvey and co-workers (5) who examined the behaviour of various enzymes to immobilised adenosine-5'-monophosphate (5'-AMP).

When polymers are prepared containing different concentrations of immobilised ligand, such that the ligand is uniformly distributed throughout the gel, the binding of an enzyme such as lactate dehydrogenase is related to the ligand concentration in a manner similar to that observed when the affinity gel is directly diluted with the underivatised Sepharose. The stronger binding observed with the heterogeneous polymers probably reflects the manner of dilution since under these conditions regions of the column would persist which still retain a ligand concentration equal to that of the original undiluted gel.

The effect of ligand concentration on the capacity of an N^6(6-aminohexyl)-5'-AMP-Sepharose column for lactate dehydrogenase using gels containing different concentrations of covalently attached ligand shows a sigmoidal response to ligand concentration. This is in contrast to the behaviour shown by the gel prepared by direct dilution of the parent gel with unmodified Sepharose (*Figure 1*). Glycerokinase is not bound to N^6-(6-aminohexyl)-5'-AMP-Sepharose at low ligand concentrations. The sigmoidal nature of the curve implies some co-

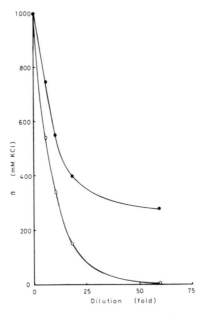

Figure 1. The binding of lactate dehydrogenase (pig heart) to N^6-(6-aminohexyl)-5'-AMP-Sepharose in relation to ligand concentration. The sample (100 μl) consisted of 5 units of enzyme and 1.5 mg of BSA. N^6-(6-aminohexyl)-5'-AMP-Sepharose (●) was diluted with unsubstituted Sepharose to the required ligand concentrations. N^6-(6-aminohexyl)-5'-AMP (○) uniformly coupled to Sepharose 4B at the indicated ligand concentration. Zero dilution is equivalent to a ligand concentration of 1.5 μmol 5'-AMP/ml of Sepharose. The binding (β) is a measure of binding strength and is the concentration of KCl (mM) needed to elute the bound enzyme. Taken from (5) with permission.

operativity in the interaction of the enzyme with the immobilised ligand and suggests that the higher ligand concentrations keep the enzyme attached to the gel by restricting its dissociation from the matrix.

Ligand concentration can be used to finely tune the binding of a macromolecule to an affinity column. The ligand concentration which results from a particular recipe is almost certainly not the one most likely to give the best separation. This is particularly true of phenylboronate columns. In these instances a range of concentrations should be tested before deciding on the final conditions for the experiment.

4. EFFECT OF COLUMN DIMENSIONS

Lowe and co-workers (6) investigated the effect of column dimension on the binding of a number of enzymes to Sepharose-immobilised 5'-AMP. When both the ligand concentration and the total amount of ligand were kept constant and the length of the gel bed altered 60-fold by a 4-fold change in diameter, the strength of the interaction was related to the column length. *Figure 2* shows how the binding of yeast alcohol dehydrogenase and glycerokinase approach a maximum limiting value as the length of a N^6-(6-aminohexyl)-5'-AMP-Sepharose column is increased.

A double-reciprocal plot of these data reveals maxima in the salt concentration required to elute the enzyme of 830 and 420 mM KCl, and column lengths for half-maximal binding of 25 mm and 15 mm, respectively, for yeast alcohol dehydrogenase and glycerokinase.

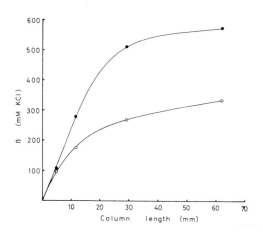

Figure 2. The effect of column length, at constant ligand concentration and content, on the binding of two enzymes to N^6-(6-aminohexyl)-5'-AMP-Sepharose. The sample (100 μl), containing enzyme (5 U) and bovine serum albumin (1.5 mg), was applied to columns of different diameters (6, 9, 13.5 and 23 mm) containing N^6-(6-aminohexyl)-5'-AMP-Sepharose (2.0 g moist weight, 1.5 μmol 5'-AMP/ ml). 'Binding' (β) refers to the KCl concentration (mM) at the centre of the enzyme peak when the enzyme is eluted with a linear gradient of KCl. Bovine serum albumin, which was eluted with the equilibration buffer, yeast alcohol dehydrogenase ($\bullet$) and glycerokinase ($\bigcirc$), were assayed by the methods cited by the authors (6).

5. FLOW-RATE AND INCUBATION TIME EFFECTS

Using the same AMP-agarose affinity adsorbent, Lowe and co-workers (6) also examined the effect of flow-rate and incubation time on the binding of lactate dehydrogenase at a low enzyme concentration (0.6 U/ml). *Figure 3* shows that under batchwise conditions there is a rapid increase in the percentage of enzyme bound during the initial time period followed by a gradual progression to 100% binding after 16 h, the half-time being 20 min.

At lower ligand concentrations a similar effect of equilibration time is observed when the adsorbent is studied under column conditions. When glycerokinase and lactate dehydrogenase are equilibrated with the adsorbent for up to 67 h both the efficiency of the column and the strength of the enzymic interaction increase with

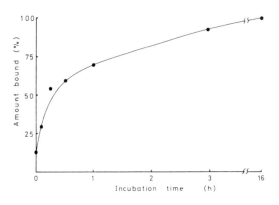

Figure 3. The effect of incubation time on the capacity of N⁶-(6-aminohexyl) 5′-AMP-Sepharose for lactate dehydrogenase under batchwise conditions. Taken from (6) with permission.

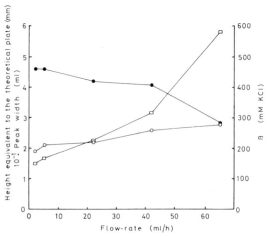

Figure 4. The binding of lactate dehydrogenase to N⁶-(6-aminohexyl) 5′-AMP-Sepharose in relation to flow-rate. The enzyme sample (100 μl) containing 5 units of lactate dehydrogenase and 1.5 mg of BSA was applied to a column (34 x 5 mm) of N⁶-(6-aminohexyl)-5′-AMP-Sepharose (0.95 g moist weight, 0.125 μmol 5′-AMP/ml). The variation of peak width (○) with flow-rate was determined at the base of the peak as was the height equivalent to the theoretical plate (□) measured as described in (6) together with β, the binding constant (●), as defined in *Figure 1*. Taken from (6) with permission.

80

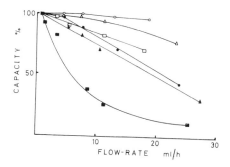

Figure 5. The dependence of capacity on flow-rate for pellicular supports *versus* agarose. Open symbols represent Matrex 201R-based ligands, closed symbols, Sepharose-based ligands. (○,●) *Lactobacillus casei* dihydrofolate reductase; (△,▲) pig heart isocitrate dehydrogenase; (■,□) *Neurospora crassa* glutamate dehydrogenase. Taken from (7) with permission.

time. In the case of glycerokinase, the percentage of enzyme bound also increases. The effect of flow-rate on the binding of lactate dehydrogenase to a column of N^6-(6-aminohexyl)-5'-AMP-Sepharose has also been examined. *Figure 4* shows that the efficiency of the column, as measured by the height equivalent to the theoretical plate, decreases at high flow-rates as does the strength of the interaction. The effect of flow-rate is more pronounced with short columns where a percentage of the enzymic activity is eluted in the void volume. No effect of inert protein concentration (bovine serum albumin) is observed even at high flow-rates (67 ml/h).

Using pellicular supports, less effect of flow-rate is observed (see *Figure 5* and ref. 7). The effect of flow-rate is also less noticeable with ligand-macromolecule interactions which have high association constants. However, with some high affinity systems (e.g., immunoglobulin-antigen), irreversible binding can occur if proteins are allowed to stay in contact with the gel for too long. Normally the column should be operated as rapidly as the K_a will allow. As mentioned above, the effect of incubation time on the adsorption process is easily studied by a batchwise procedure. *Figure 3* shows the effect of incubation time on the capacity of N^6-(6-aminohexyl)-5'-AMP-Sepharose for lactate dehydrogenase under batchwise conditions. This result can be achieved in the following experiment.

(i)   Suspend N^6-(6-aminohexyl)-5'-AMP-Sepharose (0.5 g moist weight, 1.5 μmol 5'-AMP/ml) in 10 mM potassium phosphate-KOH buffer, pH 7.5 (100 ml) containing lactate dehydrogenase (10 U pig heart muscle enzyme),

(ii)   Incubate under batchwise conditions on a roller mixer at 4.5°C.

(iii)   Stop the roller at the times indicated, allow the gel to settle (15 min) and withdraw a sample.

(iv)   Determine the activity of lactate dehydrogenase, measure the volume of the supernatant above the gel and calculate the amount bound.

Full details are described in (6).

6. pH AND TEMPERATURE EFFECTS

pH and temperature both have a pronounced effect on many affinity interactions. In cases where an enzyme has a marked pH dependence on binding (as opposed to the pH optimum activity), it can be expected that the pH profile of binding to the affinity column will follow this dependence.

Similar effects are observed with temperature. Hydrophobic interactions are decreased on lowering the temperature and the effect of hydrophobic spacer molecules on an interaction can be altered by changing temperature. If necessary subzero temperatures can be tried (8,9).

6.1 pH Effects

Early work (10,11) studying the effects of both pH and temperature dependence was carried out with group-specific adsorbents. The effect of pH on the binding of heart muscle lactate dehydrogenase to two affinity adsorbents, 6-aminohexanoyl-NAD-Sepharose and N^6-(6-aminohexyl)-5′-AMP-Sepharose, is shown in *Figure 6*. The interaction of the enzyme with both adsorbents was apparently independent of hydrogen ion concentration up to pH 8, but was dependent on the nature of the adsorbent above this pH. In the case of 6-aminohexanoyl-NAD^+-Sepharose this behaviour was characterised by an apparent pK of about 8.5, whereas with N^6-(6-aminohexyl)-5′-AMP-Sepharose the apparent pK was 9.7. The discrepancy between the apparent pK values determined with the two immobilised nucleotides is probably a reflection of the influence of the adsorbent on the interaction, since incomplete reaction of the NAD with 6-aminohexanoyl-Sepharose could result in residual charged groups on the matrix.

The percentage of glycerokinase bound to N^6-(6-aminohexyl)-5′-AMP-Sepharose decreased with pH values above 7.0; the apparent pK value was approximately 8. Furthermore, the strength of the interaction was related to the above pH profile. A similar decrease in binding was observed for the interaction of lactate dehydrogenase with 6-aminohexanoyl-NAD-Sepharose. Under identical conditions, bovine serum albumin had no affinity for this adsorbent over the

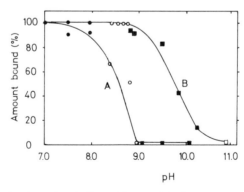

Figure 6. The effect of pH on the binding of lactate dehydrogenase to 6-aminohexanoyl-NAD-Sepharose (**A**) and N^6-(6-aminohexyl)-5′-AMP-Sepharose (**B**). Taken from (10) with permission.

pH range studied.

In dye-ligand chromatography there are several examples where raising the pH of the elution buffer weakens the binding of proteins to the column. Thus immobilised Cibacron Blue 3G-A binds plasma proteins less tightly at higher pH values (12). At pH 5.0, 95% of the proteins bind to the column, whereas at pH 9.0, only 40% bind (most of which is albumin). The pH values at which the various plasma proteins emerge from the column do not correlate with the isoelectric points of the proteins, suggesting that specific surface sites and not overall net charge are involved. This generalisation has not always been observed, however, since the binding of 6-phosphogluconate dehydrogenase from *Bacillus stearothermophilus* to Cibacron Blue-Sepharose decreases when the pH is lowered below 6.5 or increased above pH 8.5 (13).

Decreasing pH can introduce changes in the binding mechanism. Thus when using immobilised phenylboronate columns the quantitative separation of HbA$_1$c from HbA becomes difficult below pH 8.0 since a hydrophobic interaction is introduced at this point which causes the HbA to bind non-specifically to the column (14).

It is important to examine the pH dependence of binding of individual proteins to a column over as wide a range as possible in order to optimise affinity based separations.

6.2 Temperature Effects

The effect of temperature on the capacity of affinity columns is notable. Exposure of N^6-(6-aminohexyl)-5'-AMP-Sepharose to elevated temperatures reduces the percentage of yeast alcohol dehydrogenase and glycerokinase retained. However, glycerokinase can be quantitatively retained at these elevated temperatures when the concentration of the immobilised ligand is raised to 4.0 μmol 5'-AMP/ml Sepharose. *Figure 7* shows that at both ligand concentra-

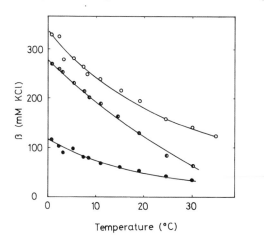

Figure 7. The effect of temperature on the binding of yeast alcohol dehydrogenase (◐,[AMP] = 1.5 μmol/ml) and glycerokinase (●,[AMP] = 1.5 μmol/ml, ○[AMP] = 4.0 μmol/ml) to N^6-(6-aminohexyl)-5'-AMP-Sepharose. Taken from (11) with permission.

83

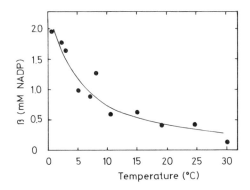

Figure 8. The influence of temperature on the binding of lactate dehydrogenase to N^6-(6-aminohexyl)-5'-AMP-Sepharose. Taken from (11) with permission.

tions the strength of the interaction (as measured by the salt concentration coincident with the peak of eluted enzyme) between the immobilised ligand and the enzymes, is reduced at increased temperatures.

Lactate dehydrogenase is so strongly bound to N^6-(6-aminohexyl)-5'AMP-Sepharose that it cannot be eluted by 1.0 M KCl at 40°C. However, elution can be effected by a linear gradient of NADH. *Figure 8* shows that the concentration of NADH required for elution of the enzyme also decreases with increasing temperature. Temperature, in general, decreases the affinity of an enzyme for the immobilised ligand.

7. IONIC STRENGTH

Electrostatic interactions are important in the binding of enzymes to affinity columns, since increasing the ionic strength (by increasing the eluting salt concentration) is usually successful in desorbing bound enzymes from many types of affinity column. For some enzymes, however, the presence of salt increases the binding, as in the hydrophobic interactions associated with the binding of fibroblast interferon (15), rabbit interferon (16) and human heart isocitrate dehydrogenase (17) to immobilised Cibacron Blue 3-GA. In such cases proteins cannot be eluted with salt and eluents such as 50% (v/v) ethylene glycol are required to desorb proteins.

The interesting effect of salt which promotes the binding of glucose-6-phosphate dehydrogenase from *Leuconostoc mesenteroides* to a column of Matrex Gel Orange B in Tris-HCl, triethanolamine/NaOH and phosphate buffers at varying ionic strengths can be shown by the following experiment (18).

(i)   Pack a series of columns (1 ml, 2.3 x 0.7 cm) with immobilised dye (ligand concentration 5.5 mg/g gel) and equilibrate with the various concentrations of buffers shown in *Table 1* (pH 7.8 at ionic strengths ranging from 10 mM to 1.0 M).

(ii) To each column, apply 100 μl (5.5 U) of purified enzyme (184 U/mg) dialysed against the appropriate buffer.

Table 1. The Effect of the Ionic Strength of Three Buffers on the Binding of Glucose-6-Phosphate Dehydrogenase to Matrex Gel Orange B. Taken from (18) with permission.

Ionic strength (mM)	Tris-HCl		Triethanolamine-NaOH		Phosphate	
	Conductivity (mS)	Bound %	Conductivity (mS)	Bound %	Conductivity (mS)	Bound %
10	0.6	0	0.75	9	1.46	89
20	0.87	53	1.38	100	2.55	100
30	1.28	89	2.00	100	3.65	100
40	1.60	100	2.65	100	4.60	100
50	1.95	100	3.25	100	6.00	100
70	2.52	100	4.40	100	7.80	100
100	3.66	100	5.80	100	10.70	100
150	5.60	100	7.80	100	15.00	100
200	6.70	100	10.80	86	19.00	100
300	11.00	93	15.50	19	28.00	100
400	12.70	69	17.50	9	35.20	100
500	14.60	14	21.20	0	41.40	100
550	16.50	0	ND	ND	ND	ND
750	19.00	0	29.50	0	53.00	87
1000	22.30	0	35.5	0	64.00	83

ND = Not determined.

(iii) Wash the columns with equilibration buffer (5 ml) and collect as one fraction.

(iv) Bound enzyme is eluted as a second fraction when the column is washed with equilibration buffer containing 1.0 M KCl (5 ml); collect as one fraction.

(v) Assay each fraction for enzyme activity and calculate the percentage enzyme bound.

8. DESORPTION METHODS

8.1 **Ligand Competition**

Most of the remarks which apply to elution by free ligand also apply to inhibitors and allosteric modifiers. In all cases it is important to achieve ligand concentrations of the column which are compatible with this technique. It is obvious that high affinity systems are not going to be easy to elute by free ligand competition if the bound ligand concentration is excessively large (see Section 3).

Elution by the inclusion of free ligand itself in the buffer is the ideal method, but can sometimes prove expensive. Often quite high concentrations of ligand have been used by experimenters; this is because the interaction with the column is so strong in the particular buffer conditions used that a high concentration of eluting ligand is needed. If the buffer is first altered to achieve a more moderate association constant, much lower concentrations of eluting ligand can be used. Careful choice of (i) the immobilised ligand concentration and (ii) selection of the appropriate concentration of eluting ligand will lead to improved and more economic use of this type of desorption method. The latter can be easily arrived at by running a linear gradient of the free ligand and measuring the concentration at the peak of the desorbed macromolecule.

To lower the ligand-macromolecule association constant before affinity elution, some knowledge of the interactions involved is needed. Increasing salt concentration may well decrease the biospecific component of binding, but it may also decrease the subsequent interaction with free ligand in affinity elution. Increased salt would minimise any non-specific ionic interactions, but it may also increase hydrophobic interactions, epecially when using polymethylene spacer arms. If the latter are very important, decreasing salt concentration may be more effective. Alternatively, a change in pH or introduction of a surface tension-reducing agent (to lessen hydrophobic and van der Waal's interactions) may be used. This could be a non-ionic detergent such as Triton X-l00, up to 1% (v/v), ethylene glycol (10 – 50% v/v) or a relatively small amount of ethanol (up to 5% v/v). The latter is unlikely to affect the value of the association constant with charged ligands. After pre-equilibration of the column with this buffer, ligand is introduced in the same buffer for the affinity elution process. It is clear that either allosteric modifying agents or inhibitors of the ligand-macromolecular interaction can also be used and that the rules governing their use follow from the above discussion.

8.2 Co-substrate Elution

Hey (18) has examined the ability of a number of biospecific eluants to desorb glucose-6-phosphate and glutamate dehydrogenases bound to immobilised triazine dyes. The data for the former enzyme are given for the purposes of illustration of the method.

Table 2 shows the results for the elution of glucose-6-phosphate dehydrogenase from *L. mesenteroides* bound to a Matrex Gel Orange B column. When biospecific elution was observed to be effective the enzyme was eluted in a sharp peak. When biospecific elution was ineffective, quantitative elution was effected by a pulse of 50 μM NADP$^+$. *Table 3* gives the purification data obtained with the effective biospecifc eluants NADP$^+$ (1.0, 0.1, and 0.05 mM), NAD$^+$ (10 mM), NADPH (1 mM) and 2',5'-ADP (1 mM). Of the many eluants tested only a few were observed to be effective. Thus both of the coenzymes NAD$^+$ and NADP$^+$ utilised by this dual nucleotide-specific enzyme were capable of desorbing purified enzyme as well as NADPH and 2',5'-ADP. The substrate and product of the reaction catalysed by the enzyme, glucose-6-phosphate and 6-phosphogluconate, respectively, were ineffective at the concentrations tested.

8.3 Ionic Strength Alterations

Historically, elution has been most frequently effected by simply increasing the ionic strength of the column wash buffer. The economy of such a method outweighs any other considerations. Many papers describe the use of either KCl or NaCl for increasing the ionic strength, but the reader should refer to the work of Robinson *et al.* (19) on the application of B coefficients of viscosity to predict elution by different salts of dye-ligand and presumably other affinity columns.

Table 2. A Study of a Range of Biospecific Eluants of Glucose-6-Phosphate Dehydrogenase from *L. mesenteroides* Bound to an Immobilised Matrex Gel Orange B.

Biospecific eluant	Concentration (mM)	Amount of enzyme eluted (% of that applied)
NADP$^+$	1	85
NADP$^+$	0.1	80
NADP$^+$	0.05	78
NADP$^+$	0.01	13
NAD$^+$	1	11
NAD$^+$	10	74
NADPH	1	83
NADH	1	4
Glucose-6-phosphate	1	7
Tetra-sodium pyrophosphate	1	0
Tetra-sodium pyrophosphate	10	0
6-phosphogluconate	1	0
2′,5′-ADP	1	95
2′,5′-ADP	0.05	6
3′,5′-ADP	1	0
2′-AMP	1	0
3′-AMP	1	0
5′-AMP	1	0
5′-ADP	1	0
5′-ATP	1	3
Nicotinamide mononucleotide	1	0
Pyridoxal 5′-phosphate	1	5
Glucose-6-phosphate + NADH	1	8
6-phosphogluconate + NADH	1	0
Tetra-sodium pyrophosphate + NADH	1	0
Glucose-6-phosphate + NAD$^+$	1	0
Pyridoxal-5′ + NAD$^+$	1	14

Table 3. Summary of the Effective Biospecific Elution of Glucose-6-Phosphate Dehydrogenase from *L. mesenteroides* from Procion Red HE-3B and Matrex Gel Orange B Cibacron Blue 3-GA Columns.

Eluant	Procion Red HE-3B		Cibacron Blue 3-GA		Matrex Gel Orange	
	Purifn. (-fold)	Yield (%)	Purifn. (-fold)	Yield (%)	Purifn. (-fold)	Yield (%)
1 mM NADP$^+$	16	73	15	82	28.2	85
100 µM NADP$^+$	–	–	–	–	29.9	80
50 µM NADP$^+$	11	59	11	19	40.3	78
1 mM NADPH	12	70	11	67	38.4	83
10 mM NAD$^+$	–	–	–	–	39.1	74
1 mM 2′,5′-ADP	8	63	9	70	21.6	95
1 mM G-6-P + 1 mM NADH	5	30	5	28	–	–

Using the published tables, it is possible to alter the ionic strength concentration coincident with the peak of a protein's elution and also the elution profile of a column by varying the salt composition and concentration.

8.4 Solvent Changes

Binding of macromolecules can be affected by solvent changes. For example, ethylene glycol and other alcohols, especially when combined with low temperatures (Section 6), have all been used in varying conditions (see also Section 8.9). An example of the use of ethylene glycol is as follows.

(i)   Equilibrate a column (2 ml) of immobilised Procion Yellow MX-3R (ligand concentration 17.2 mg/g gel) in 50 mM Tris-HCl buffer, pH 7.8.
(ii)   Apply a sample (1 ml, 32.2 U, 8.8 mg,) of partially purified glucose-6-phosphate dehydrogenase.
(iii) Wash the column with equilibration buffer at a flow-rate of 4 ml/h (1 ml fractions) until the A_{280} of the washings drops below 0.05.
(iv)   Apply a pulse (10 ml) of 50% (v/v) ethylene glycol in equilibration buffer followed by a further wash with equilibration buffer.
(v)   Monitor all fractions at 280 nm and assay for enzyme activity.
(vi)   Pool and determine the yield and degree of purification.

8.5 Temperature Elution

Harvey and co-workers (11) showed that the effects of temperature can be utilised to elute enzymes from affinity adsorbents. *Figure 9* illustrates the resolution of yeast alcohol dehydrogenase, glycerokinase, hexokinase and lactate dehydrogenase on immobilised 5′-AMP using a linear temperature gradient. It is relevant to note that glycerokinase and yeast alcohol dehydrogenase are eluted in the order expected from their apparent energies of adsorption with almost quantitative recovery (70–90%) whilst lactate dehydrogenase even at 40°C still requires a pulse of 5 mM NADH for elution.

Temperature, when used in combination with other techniques, such as lower ligand concentrations, NADH pulses or ionic strength gradients, can thus be used to effect the elution of tightly bound enzymes. Furthermore, temperature gradients can provide a very sensitive means of eluting weakly bound enzymes. In addition, eluted enzymes which are uncontaminated with desorbing species (e.g., salt or nucleotide pulses) can be employed directly for kinetic studies.

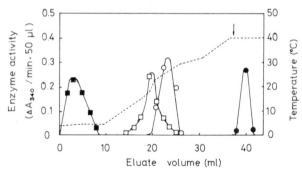

Figure 9. The separation of various enzymes on N^6-(6-aminohexyl)-5′-AMP-Sepharose using an increasing temperature gradient. (■)hexokinase, (□)glycerokinase, (○) yeast alcohol dehydrogenase, (●) lactate dehydrogenase H_4. The arrow indicates elution with a pulse of 5 mM NADH. Taken from (11) with permission.

8.6 Buffer and/or pH changes

Sometimes surprising results can occur which can be put to good effect in affinity chromatography. The serendipitous discovery that certain buffer ions are required for the binding of glucose-6-phosphate dehydrogenase to Matrex Gel Orange A led to the following eminently scaleable method for the production of this enzyme.

(i) Equilibrate a column (2 ml, 3.0 x 0.8 cm) of Matrex Gel Orange B (ligand concentration 17.2 mg/g gel) with 50 mM Tris-HCl buffer, pH 7.8.

(ii) Apply a sample of partially purified *L. mesenteroides* glucose-6-phosphate dehydrogenase (by previously passing crude bacterial cytosol through a column of Matrex Gel Purple A)(40 U, 7.3 mg, sp. act. 5.5 U/mg), to the column.

(iii) Wash the column with equilibration buffer (12 ml) at a flow rate of 4 ml/h (1 ml fractions) until the absorbance at 280 nm drops to below 0.05.

(iv) Apply a retro-gradient (30 ml) of Tris-HCl buffer, pH 7.8 from 50 to 0 mM. Make the gradient in the usual apparatus but remember to reverse the contents of each container so that the 50 mM buffer is the closer of the two to the column inlet.

(v) Monitor all fractions for absorbance at 280 nm and assay for enzyme activity and protein to enable the yield and degree of purification to be determined.

Typical results are shown in *Figure 10*.

8.7 Chaotropic Reagents

Table 4 lists some of the reagents used to desorb antibody-antigen complexes when one of the two components is immobilised. Many of these reagents are chaotropic. Sodium dodecyl sulphate, urea or guanidinium salts, iodide or thiocyanate can all be used in the concentrations given in the table.

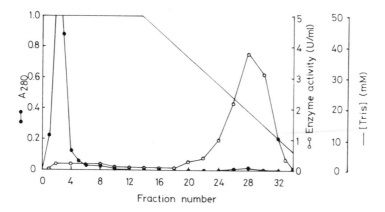

Figure 10. The chromatography of *L. mesenteroides* glucose-6-phosphate dehydrogenase on immobilised Matrex Gel Orange B using a Tris-HCl retro-gradient (18).

Table 4. Some Eluants Suitable for Use with Immunoadsorbents. From Lowe and Dean (20) with permission.

Hapten	*Eluant*	*Reference*
(i) Acids (pH 3)		
Insulin	0.01 – 1 M HCl	
Insulin	1 M Acetic acid, pH 2.0	21
Anti-DNP-amino acids	20% (v/v) Formic acid	22
	Glycine-HCl, pH 2.8	23
Ragweed allergens	Glycine-H_2SO_4, pH 2.7	24
	1 M Propionic acid	25
(ii) Salts		
IgG	0.2 – 2 M sodium thiocyanate	26
	2.5 M NaI, pH 7.5	27
	2.8 M $MgCl_2$	28
Semliki Forest virus	Carbonate/bicarbonate, pH 11	29
(iii) Protein denaturants		
SDS	0.01 – 0.1% (w/v)	
Human chorionic somatomammotropin	6 M Guanidine-HCl, pH 3.1	30
β-Lipoprotein	4 M Urea	31
Galactosidase	8 M Urea	32
(iv) Haptens		
p-Aminophenyllactoside	0.5 M Lactose	33
DNP-Lysine	DNP, DNP-lysine	34
(v) Others		
Human complement, C1	0.2 M 1,4-Diaminobutane	35
IgG	Distilled water	36

8.8 Electrophoretic Desorption

Electrophoretic desorption has been developed as a mild, non-chaotropic technique for the removal of charged material from affinity matrices and in particular immunoadsorbents (37,38). The technique is capable of excellent yields and is applicable to all situations where a reversible equilibrium describes the equilibrium process. It obviates the need for competing ligand elution and also for concentration of large volumes of eluate. The technique has been applied to a variety of interactions: anti-steroid antibody/immobilised steroid (39,40), immunoglobulin/immobilised protein A (41), anti-Concanavalin A (Con A) second antibody/immobilised Con A (41), ferritin/immobilised anti-ferritin antibody (42), sex hormone binding globulin/immobilised androstanediol (43) and human serum albumin/immobilised Cibacron Blue 3G-A(44).

Electrophoretic desorption of protein from a matrix may be carried out using a modified Macro polyacrylamide gel electrophoresis cell (Birchover Instruments Ltd., Letchworth, UK) The desorption cell is shown in *Figure 11*. The polyacrylamide gel (7.5%) is cast with a lip on the upper surface, acting to prevent loss of matrix between the sides of the gel and the gel holder. A disc of

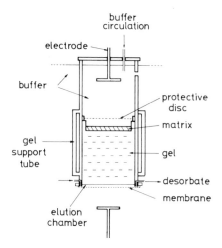

Figure 11. The desorption cell for removing bound macromolecules from affinity gels with electric fields. Taken from (44) with permission.

porous polypropylene resting on the gel lip prevents matrix disturbance caused by the high rates of buffer circulation (500 ml/min). This buffer circulation prevents any possible local pH changes. Gels are pre-electrophoresed for 30 min prior to layering the matrix onto the gel surface. The flow of current should be such that, at the pH used, the protein migrates through the gel towards the elution chamber.

The instrument utilises a 12-channel punched-card programme to control the power supply, periodic evacuation of the desorption chamber and time between such evacuations automatically. The buffer temperature during electrophoresis is controlled by means of a cooling coil submerged in the buffer reservoir connected to a thermocirculator. The protein concentrations in the fractions obtained from successive evacuations of the desorption chamber (the desorbate) may then be determined.

Good recoveries can be obtained by optimising the above parameters. However, an alternative method has been described by Haff *et al.* (45) which uses isoelectric focusing to dissociate the bound material from the matrix. The affinity gel is placed in a pre-focused pH gradient, which is produced by electrophoresis of a flat bed ampholyte-containing gel. The system operates at low current and high voltage. Although neither this method nor electrophoretic desorption has really been optimised, both are capable of dissociating high affinity interactions such as biotin-avidin and immunoadsorbents under mild conditions.

8.8.1 *Separation of IgG from Immobilised Antigens*

(i)   Prepare a flat bed isoelectric focusing gel from G-200 Superfine Sephadex beads pre-swollen in a suitable buffer and containing $1-5\%$ (v/v) carrier ampholyte solution (the concentration of ampholyte solution varies with the manufacturer — follow the instructions from the supplier).

(ii)   In the first instance, use broad range carrier ampholytes (pH $3-10$).

(iii) Place an electric field across the bed and allow the bed to focus for 2000 V/h with cooling.

(iv) When the pH gradient has been established, switch off the power supply, place the affinity gel into a slot which is at least $1-2$ pH units away from the isoelectric point of the protein (in the case of IgG choose pH values of 5.5/6.0 or 10).

(v) Reconnect the power supply and focus at a voltage which is compatible with the cooling system.

(vi) Over the next 1 h remove G-100 in the region of the pI of IgG and test for antibody.

Pharmacia supply the equipment for this application and their literature and reference 45 should be consulted before carrying out this experiment.

9. USE OF CYLINDRICAL BAFFLES IN LARGE-SCALE AFFINITY COLUMNS

9.1 Introduction

Affinity chromatography has been widely used for the purification of biological materials on a small scale. Large diameter chromatography columns are normally needed for industrial applications. In large-scale columns, however, compaction of the adsorbent bed limits the maximum superficial liquid velocity and tends to cause non-uniform flow of liquid through the bed. Sada *et al.* (46) have studied the effect of insertion of coaxial cylindrical baffles in a column of 10.0 cm (i.d.) on the maximum superficial velocity, breakthrough and elution profiles. A summary of these authors' experiments is given in the following section.

9.1.1 *Comparison of Chromatography with and without Baffles*

(i) Prepare the adsorbent by the methods described in Chapter 1 by immobilising soybean trypsin inhibitor to CNBr-activated Sepharose 4B.

(ii) Mix the adsorbent with Sepharose 4B in the ratio of 7:93 (v/v) in order to adjust the adsorption capacity of the bed.

(iii) Use two acrylic resin columns of 1.44 and 10.0 cm (i.d.).

(iv) At the top and bottom of each column, place sintered glass filters to produce uniform distribution of liquid flow.

(v) Construct one of two types of coaxial baffles, one consisting of three coaxial cylinders of 2.5, 5.0 and 7.5 cm (i.d.) and the other a cylinder of 5.0 cm i.d. (both made of stainless steel with 0.4 mm thickness) and insert into the 10.0 cm column.

(vi) Pack each column with an equal amount of the adsorbent per unit cross-sectional area of the column to observe the effects of the column diameter and the inserted baffles on:
    (a) the pressure drop through the bed,
    (b) the breakthrough profiles and
    (c) elution curves.

(vii) Equilibrate the column with buffer solution (0.05 M Tris-HCl, containing 0.02 M $CaCl_2$, pH 8.2).

(viii) Dissolve trypsin in the same buffer solution and pump onto the column with a peristaltic pump.

(ix) Wash the column with the same buffer, and elute the trypsin with a solution containing 0.005 M HCl and 0.5 M NaCl.

(x) Measure the absorbance of the effluent solution at 280 nm spectrophotometrically.

(xi) Measure the pressure drop through the bed with a pressure gauge. Carry out all measurements at 20°C.

(xii) Determine the activity and purity of trypsin using benzoyl arginine *p*-nitroanilide HCl and *p*-nitrophenyl *p'*-guanidino benzoate HCl.

9.2 Results and Discussion

9.2.1 *Maximum Superficial Velocity*

In *Figure 12* the height of the adsorbent bed and the pressure drop are plotted against the superficial liquid velocity for the column of 1.44 and 10.0 cm (i.d.) with and without the cylindrical baffles of 9 cm height. The pressure drop increased and the bed height decreased as the liquid velocity increased. When a certain liquid velocity was reached, the pressure drop rapidly increased because of compaction of the Sepharose beads. The maximum velocity of the 1.44 cm column was higher than that of the 10.0 cm column without the baffles and almost equal to that with the baffle of 2.5, 5.0 and 7.5 cm (i.d.).

The effects of insertion of the baffles on elution profiles were remarkable (see *Figure 13*). The elution profile for the 10.0 cm column with the baffle of 2.5, 5.0 and 7.5 cm (i.d.) coincided with that for the 1.44 cm column. On the other hand, elution of trypsin was prolonged for the 10.0 cm column without the baffle.

Increase of the diameter of unmodified columns results in a decrease in the maximum liquid velocity and the profiles of the breakthrough and elution curves

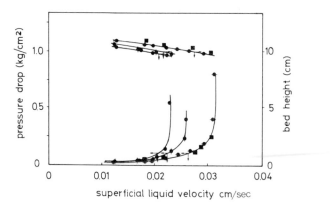

Figure 12. The variation of pressure drop and bed height with superficial liquid velocity for two diameters of column and with different baffles. Diameters were as follows: (■) 1.44 cm column without baffles, (●) 10.0 cm column without baffle, (♦) 10.0 cm column with 5.0 cm baffle (◕), 10.0 cm column with 2.5, 5.0, 7.5 cm baffle. Taken from (46) with permission.

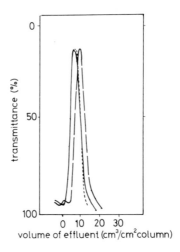

Figure 13. Elution curves with and without the baffle (bed height of ~10 cm, liquid velocity = 0.02 cm/sec) with diameter as follows: (——) 1.44 cm without baffle, (····) 10 cm with 2.5, 5 and 7.5 baffle, (— —) 10 cm without baffle. Taken from (47) with permission.

are broader because of the compacting of the Sepharose beads. These disadvantages, however, are avoided by insertion of coaxial cylindrical baffles in the column.

10. DESIGN AND USE OF HIGH PERFORMANCE AFFINITY CHROMATOGRAPHY MINICOLUMNS

by R.R.Walters

Since affinity chromatography is so selective, very short colums can provide excellent separations. Although disposable pipettes or other makeshift columns are suitable for simple separations, for rapid and sensitive analytical work special high-performance affinity chromatography (h.p.a.c.) columns which minimise band broadening and can withstand moderate pressures are necessary.

10.1 Design

The column fittings are designed for use with 0.25 in (o.d.) steel column tubes and 0.0625 in (o.d.) connecting tubing. Columns ranging from 2 mm to 10 cm in length have been prepared in our laboratory. The end fittings can be made from steel or Delrin. The latter is easier to machine and can also withstand pressures of up to 5000 p.s.i.

Details of the column design are shown in *Figure 14*. The inlet and outlet end fittings (A) are machined from 0.75 in hexagonal type 304 stainless steel or Delrin. Connections to inlet and outlet tubing are made to accept 10/32 threaded 0.0625 in fittings (B), such as Swagelok or Parker fittings. A 0.0625 in diameter hole (C) connects to a 0.0135 in hole (D) with a tapered opening of 0.08 in diameter and 0.015 in depth (D). A channel (I) of depth 0.160 in aids in removal of frits (E) and O-rings (H).

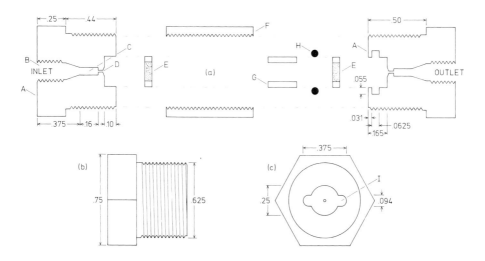

Figure 14. Construction of column fitting: (**a**) exploded longitudinal cross-section of the end fittings, column and frits; (**b**) side view of the inlet end fitting; (**c**) end view from inside the column of the inlet and outlet end fittings. All dimensions are in inches. Dimensions of the inlet and outlet end fittings are the same except as indicated. Other details are given in the text.

The end fittings are connected together by a 0.75 in, hexagonal brass connector (F) of length approximately 0.5 in greater than the length of the column blank (G). The brass connector and end fittings have 0.625 in - 32 threads. The 0.25 in (o.d.) type 316 low RMS stainless steel column blank (G) (Alltech, Deerfield, IL) is cut and deburred on a lathe.

Seals are formed via two 0.25 in (o.d.) fluorocarbon-encased stainless steel frits (E) (Alltech) whose frit diameters match the internal diameter of the column blank. A 0.25 in (i.d.) x 0.0625 in rubber O-ring (H) fits into a groove in the outlet end fitting. The O-ring holds the column blank in the outlet end fitting but is not part of the high-pressure seal.

The brass connector (F) can be used as a water jacket by attaching copper tubes 0.5 in from either end of the connector. A sealant must also be applied to the threads of the connector.

10.2 Column Packing

Particles of 10 μm and smaller diameter are usually packed using high-pressure slurry methods (47,48). The necessary equipment is not found in many biochemical laboratories, therefore we describe two alternative methods which yield short columns with efficiencies comparable with those of slurry-packed columns.

In any column packing method it is important to remove fines and large particles from the packing. Fines can be removed by suspending the packing in buffer, sonicating briefly, and pouring off the fines after the bulk of the packing has settled. Large particles, such as agglomerates of the packing and dust, can be removed either by removing the material which first settles out or by pouring the packing through a fine sieve (40 μm).

The amount of packing needed can be estimated using a packing density of $0.4-0.6$ g/ml for silica packings.

When using slurry packing methods, columns are much more efficient when packed using an aqueous buffer rather than pure water.

10.2.1 *Dry Packing*

(i)   Fit an outlet end fitting (A) with a frit (E) and O-ring (H).
(ii)   Press a column tube (G) into the end fitting.
(iii) Add the dry packing to the column in small increments, tapping the column on the benchtop a few times between each increment.
(iv)   When filled, level off with the edge of a spatula.
(v)   Finish assembling the column using an inlet end fitting (A), frit (E) and connector (F).

10.2.2 *Vacuum Slurry Packing*

This method is more convenient to use than dry packing since the packing does not have to be dried after immobilising the affinity ligand. The packing is suspended as a concentrated slurry and continuously pipetted into an outlet end fitting (assembled as above) which is connected to an aspirator vacuum. Only about 1 min is needed to pack the column. The packing is then levelled off and the column assembled as above.

10.2.3 *Sample Size*

The maximum amount of sample which can be applied to the column without some of it eluting in the void volume (the dynamic adsorption capacity) depends on several parameters: the surface area of the support, the surface concentration and activity of immobilised ligand, and the kinetic and equilibrium constants of the system. For 6 mm minicolumns, we have found dynamic capacities in the range of $5-500$ μg, or approximately $20-200$ μg/m^2 of surface area (49).

10.2.4 *Extracolumn Effects*

The bands which elute from short columns are in a small volume. The volumes of the connecting tubing, heat exchanger tubing, and flow cell can all cause dilution and broadening of the bands (49). For the fastest and most sensitive separations, these volumes should be minimised. H.p.l.c. equipment which has been specially designed for microbore h.p.l.c. columns is the most suitable. If elution is to be accomplished using step or rapid gradient mobile phase changes, the h.p.l.c. mixing chamber should be of very small volume or else replaced with a T-connector. When the proper equipment is used, h.p.a.c. separations can be repeated at intervals of less than 1 min.

11. REFERENCES

1. Falkbring,S.O., Gothe,P.O., Nyman,P.O., Sundberg,L. and Porath,J. (1972) *FEBS Lett.*, **24**, 229.
2. Jackson,R.J.,Wolcott,R.M. and Shiota,T. (1973) *Biochem. Biophys. Res. Commun.*, **51**, 428.
3. Tesser,G.I., Fisch,H.U. and Schwyzer,R. (1974) *Helv. Chim. Acta*, **57**, 1718.
4. Inman,J.K. and Dintzis,H.M. (1969) *Biochemistry (Wash.)*, **8**, 4074.
5. Harvey,M.J., Lowe,C.R., Craven,D.B. and Dean,P.D.G. (1974) *Eur. J. Biochem.*, **41**, 335
6. Lowe,C.R., Harvey,M.J. and Dean,P.D.G. (1974) *Eur. J. Biochem.*, **41**, 341.
7. Dean,P.D.G. and Watson,D.H. (1978) in *Affinity Chromatography*, Hofmann-Ostenhoff,O. *et al.* (eds.), Pergamon Press, Oxford, p.25.
8. Balny,C. and Douzou,P. (1979) *Biochimie*, **61**, 445.
9. Douzou,P. (1979) *Q. Rev. Biophys.*, **12**, 521.
10. Lowe,C.R., Harvey,M.J. and Dean,P.D.G. (1974) *Eur. J. Biochem.*, **41**, 347.
11. Harvey,M.J., Lowe,C.R. and Dean,P.D.G. (1974) *Eur. J. Biochem.*, **41**, 353.
12. Angal,S. and Dean,P.D.G. (1977) *Biochem. J.*, **167**, 301.
13. Qadri,F. and Dean,P.D.G. (1980) *Biochem. J.*, **191**, 53.
14. Middle,F.A. and Dean,P.D.G. unpublished observations
15. Jankowski,W.J., von Neuenchhausen,W., Sulkowski,E. and Carter,W.A. (1976) *Biochemistry (Wash.)*, **15**, 5182.
16. Bollin,E.J.R., Vastola,K., Oleszek,D. and Sulkowski,E. (1978) *Prep. Biochem.*, **8**, 259.
17. Seelig,G.F. and Colman,R.F. (1977) *J. Biol. Chem.*, **252**, 3671.
18. Hey,Y. (1983) Ph.D. Thesis, Liverpool University.
19. Robinson,J.B.,Jr., Strottman,J.M. and Stellwagen,E. (1981) *Proc. Natl. Acad. Sci., USA*, **78**, 2287.
20. Lowe,C.R. and Dean,P.D.G (1974) *Affinity Chromatography*, published by Wiley, London.
21. Cuatrecasas,P. (1969) *Biochem. Biophys. Res. Commun.*, **35**, 531.
22. Akanuma,Y., Kuzuya,T., Hayashi,M., Ide,T., and Kuzuya,N. (1970) *Biochem. Biophys. Res. Commun.*, **38**, 947.
23. Givol,D., Weinstein,Y., Gorecki,M. and Wilchek,M. (1970) *Biochem. Biophys. Res. Commun.*, **38**, 947.
24. Ternynck,T. and Avrameas,S. (1971) *Biochem. J.*, **125**, 297.
25. Spitzer,R.H., Kaplan,M.A. and Leija,J.G. (1968) *Int. Arch. Allergy*, **34**, 488.
26. Tanigaki,N., Kitagawa,M., Yagi,Y. and Pressman,D. (1967) *Biochem. Biophys. Res. Commun.*, **27**, 747.
27. Dandliker,W.B., Alonso,R., de Saussure,U.A., Kierszenbaum,F., Levison,S.A. and Schapiro,H.C. (1967) *Biochemistry (Wash.)*, **6**, 1460.
28. Avrameas,S. and Ternynck,T. (1967) *Biochem. J.*, **102**, 37C.
29. Avrameas,S. and Ternynck,T. (1969) *Immunochemistry*, **6**, 53.
30. Wood,K.R., Stephen,J. and Smith,H. (1968) *J. Gen. Virol.*, **2**, 313.
31. Mougdal,N.R. and Porter,R.R. (1963) *Biochim. Biophys. Acta*, **71**, 185.
32. Weintraub,B.D. (1970) *Biochem. Biophys. Res. Commun.*, **39**, 83.
33. Beaumont,J.L. and Delphanque,B. (1969) *Immunochemistry*, **6**, 489.
34. Melchers,F. and Messer,W. (1970) *Eur. J. Biochem.*, **17**, 267.
35. Wofsy,L. and Burr,B. (1969) *J. Immunol.*, **103**, 380.
36. Bing,D.H. (1971) *J. Immunol.*, **107**, 1243.
37. Morgan,M.R.A., Brown,P.J., Leyland,M.J. and Dean,P.D.G. (1978) *FEBS Lett.*, **37**, 239.
38. Morgan,M.R.A. and Dean,P.D.G. (1978) in *Theory and Practice in Affinity Techniques*, Sundaram,P.V. and Eckstein,F. (eds.), Academic Press, p.15.
39. Grenot,C. and Cuilleron,C. (1977) *Biochem. Biophys. Res. Commun.*, **79**, 274.
40. Morgan,M.R.A., Kerr,E.J. and Dean,P.D.G. (1978) *J. Steroid Biochem.*, **9**, 767.
41. Morgan,M.R.A., Johnson,P.M. and Dean,P.D.G. (1978) *J. Immunol. Methods*, **23**, 381.
42. Brown,P.J., Leyland,M.J., Keenan,J.P. and Dean,P.D.G. (1977) *FEBS Lett.*, **83**, 256.
43. Iqbal,M.J., Ford,P. and Johnson,M.W. (1978) *FEBS Lett*, **87**, 235.
44. Morgan,M.R.A. Slater,N.A. and Dean,P.D.G. (1978) *Anal. Biochem.*, **92**, 144.
45. Haff,L.A., Lasky,M. and Manrique,A. (1979) *J. Biochem. Biophys. Methods*, **1**, 275.
46. Sada,E, Katoh,S. and Shiozawa,M. (1982) *Biotechnol. Bioeng.*, **24**, 2279.
47. Walters,R.R. (1983) *Anal. Chem.*, **55**, 591.
48. Snyder,L.R. and Kirkland,J.J. (1979) *Introduction to Modern Liquid Chromatography*, published by Wiley, New York, p.216.
49. Walters,R.R. (1983) *Anal. Chem.*, **55**, 1395.

CHAPTER 5

Ligands for Immobilisation

1. NUCLEIC ACIDS

1.1 **Binding of DNA to CM-cellulose**

1.1.1 *Materials*

Carboxymethyl-cellulose (CM-cellulose) of the type CM-23 No 45030 may be bought from SERVA Feinbiochemical GmbH, Heidelberg. Calf thymus DNA, highly polymerised from Sigma, St.Louis, USA.

1.1.2 *Preparation of the Column Material (Standard Procedure)*

(i)   Suspend CM-cellulose (1 g) in 0.5 M NaOH (50 ml) and stir for 30 min.
(ii)   After decantation from the supernatant, wash the material with water until the effluent reaches a pH of 8.0.
(iii)   Suspend the CM-cellulose derivative in water (100 ml) and adjust the pH to 3.5 by addition of 0.01 M HCl.
(iv)   Filter under suction, wash the material with ethanol/ether 1:1 (v/v) (3 x 5 ml), followed by ether (20 ml) and finally dry for 60 min at 40°C.
(v)   To the prepared CM-cellulose add calf thymus DNA (20 mg) dissolved in water (10 ml).
(vi)   Spread the resulting suspension over the surface of a Petri dish and dry slowly over a period of 60 h at 40°C, controlling the drying process by periodically covering the dish.
(vii)   Scrape the resulting DNA-cellulose from the surface of the dish, pulverise, suspend in 0.05 M phosphate buffer, pH 7.0 containing 50% (v/v) glycerol (50 ml) and leave for 24 h at room temperature.
(viii) Wash the material with water (10 x 10 ml) and store at 4°C in 1.0 M NaCl solution.

If the recommended starting cellulose is used, the resulting column material contains about 15 mg DNA per gram of dry CM-cellulose which corresponds to about 2.5 mg/ml bed volume.

1.2 **Immobilisation of Nucleic Acids to Bisoxirane-activated Polysaccharides**

1.2.1 *Activation of Matrices*

Bis-oxirane activation of Sepharose 6B is carried out according to Porath and Sundberg (1).

(i)   Activate cellulose (10 g) by suspending in 1.0 M NaOH (30 ml) containing sodium borohydride (60 mg).
(ii)   Add 1,4-butanediol diglycidylether (20 ml) and shake the mixture vigorously for 20 h at room temperature.

(iii) After filtration under suction wash the cellulose extensively with ice-cold distilled water until the effluent is neutral.

(iv) After lyophilisation, store the material desiccated at $-20°C$. Cellulose matrices activated in this way contain approximately $10-15$ μmol of reactive groups/g.

1.2.2 *Immobilisation of Polynucleotide*

(i) Dissolve the desired amount of the polynucleotide in 20 mM KCl $(5-10$ ml/g of dry matrix) containing 10% (v/v) 1,4-dioxane, and adjust the pH of the solution to 12.0 by addition of triethylamine.

(ii) After adding to the activated matrix spread the mixture over the surface of a Petri dish and allow to evaporate slowly over a period of approximately 60 h at 35°C in an atmosphere of about 30% relative humidity.

(iii) In order to inactivate unreacted epoxy groups, resuspend the material in a small amount of 0.01 M NaOH and leave to stand at room temperature for 5 days. This method is not suitable for polyribonucleotides.

1.2.3 *Desorption of Unreacted Ligand*

Since nucleic acids and polynucleotides of high molecular weight interact non-specifically with polysaccharide matrices, it is essential to remove the unreacted ligand by washing the material thoroughly in a desorbing solution. Suspend the adsorbent in a solution of low ionic strength (preferably distilled water) containing $20-50\%$ (v/v) of a hydrophilic organic solvent such as glycerol or formamide and agitate gently for 2 days. Wash the product with distilled water and store in any neutral buffer solution (preferably citrate) or in lyophilised form at $0-5°C$.

1.3 **Immobilisation of DNA to Cellulose Using a Water-soluble Carbodiimide as the Coupling Agent**

(i) Wash the cellulose before coupling, by suspending cellulose in 1% (v/v) methanolic-HCl for 3 days at room temperature followed by extensive washing of the cellulose with water on a sintered glass funnel, then air-dry.

(ii) Dissolve the DNA or other polynucleotide in water, and adjust the pH of the solution to 6.0 with dilute HCl or NaOH.

(iii) Add this solution, containing $2.5-250$ μmol of DNA/ml, to one-quarter the volume of 0.2 M MES buffer, pH 6.0 containing water-soluble carbodiimide (final concentration 100 mg/ml). [Either N-cyclohexyl-N$'$-β -(γ-methylmorpholinium) ethyl carbodiimide *p*-toluene sulphonate or 1-ethyl-3(3-dimethylamino propyl)-carbodiimide hydrochloride are suitable water-soluble coupling agents].

(iv) Add the buffer-carbodiimide mixture to the clean, dry cellulose powder making a thick paste.

(v) Spread out the paste in a Petri dish and allow to dry slowly at room temperature.

(vi) After 2 h, 5 h, 8 h and 14 h, place the Petri dish in a water-saturated atmosphere in a closed tank for 1 h and then return to the normal atmosphere.

(vii) After 24 h, wash the dry cellulose powder extensively by agitation in a large volume of 0.05 M potassium phosphate buffer, pH 7.0, and collect by filtration on a sintered glass funnel.

(viii) Repeat this procedure three to five times, and finally wash with water.

(ix) Collect the DNA-cellulose by filtration and air-dry. Alternatively, the cellulose may be suspended in 0.1% (w/v) ammonium hydroxide between the first and second potassium phosphate buffer washes. This procedure removes small amounts of carbodiimide that may be present as substituents on the uracil, thymine, or guanine bases.

1.4 Preparation of Oligo(dT)-cellulose

(i)   Add 2 μmol of thymidine-5'-phosphate (as the pyridinium salt) in dry pyridine (3 ml) to a 2-fold excess of dicyclohexylcarbodiimide.

(ii) Add a few glass beads and shake the thick gum that forms for 5 days at room temperature.

(iii) Then add washed cellulose (5 g), dried at 100°C for 10 h, to the mixture with dry pyridine (50 ml) and an additional 2 mmol of dicyclohexylcarbodiimide.

(iv) Shake this mixture for a further 5 days at room temperature, then collect the cellulose by filtration and wash with pyridine.

(v)   Suspend the washed oligo(dT)-cellulose in 50% (v/v) aqueous pyridine for 10 h, collect by centrifugation, and then wash with warm ethanol to remove all of the dicyclohexylurea that accumulates during the synthesis.

(vi) Finally, wash the product by filtration with distilled water until the absorbance of the washings at 267 nm is close to zero.

A similar oligo(dT)-cellulose preparation can be made in aqueous media using water-soluble carbodiimides. The coupling of thymidine-5'-phosphate is nearly identical ($\sim$60%) to that achieved by the non-aqueous method. The aqueous method is essentially the same as that described for the attachment of DNA to cellulose by way of carbodiimide-mediated coupling; however, thymidine-5'-phosphate in MES buffer, pH 6.0, is used instead of aqueous DNA solutions.

1.5 Preparation of Poly(U)-cellulose

One of the easiest of all affinity hybridisation matrices to synthesise is poly(U)-cellulose.

(i)   Add washed cellulose (1 g) to a solution of commercial poly(U) (8 mg/ml) in distilled water.

(ii) Mix the cellulose and poly(U) solutions thoroughly, lyophilise the product and resuspend in 95% (v/v) ethanol (10 ml).

(iii) Pour the suspension into a Petri dish and irradiate for 15 min with a 30 W u.v. lamp held 20 cm from the surface. The irradiation creates cross-links between the poly(U) and the cellulose matrix.

(iv) Wash the irradiated poly(U)-cellulose with distilled water, resuspend and store in 0.01 M Tris-HCl, pH 7.5, containing 0.9 M NaCl.

Although only about 5% (0.4 mg) of the initial poly(U) is coupled to the cellulose, the binding capacity of poly(A) to a column of poly(U) cellulose is excellent. A column (0.8 cm x 20 cm) of this material will retain at least 1.5 mg of poly(A).

1.6 Preparation of Poly(U) Coupled to Glass Fibre Filters

(i)   Apply a solution of poly(U) (1 mg/ml) in distilled water to 2.4 cm diameter fibreglass filters (0.15 ml to each filter).

(ii)   Air-dry the filters at 37°C and then irradiate on each side for 2.5 min under the same conditions employed above for poly(U)-cellulose.

(iii) Wash the filters with water and dry.

About 0.1 mg of poly(U) is bound to each filter under these conditions.

1.7 Preparation of Poly(U) Coupled to CNBr-activated Agarose

(i)   Wash agarose (4%) beads (40 g wet weight) and suspend in ice-cold water (60 ml).

(ii)   Add cyanogen bromide (9 g) dissolved in water (135 ml) and adjust the pH to 11.5. Maintain this pH by the addition of concentrated KOH as necessary.

(iii) After 20 min collect the agarose by filtration and wash with ice-cold 0.1 M potassium phosphate buffer, pH 8.0 (600 ml).

(iv)   Resuspend the washed agarose in 40 ml of the same buffer containing poly(U) (20 mg).

(v)   Incubate the mixture at 4°C for 18 h and then wash with 90% (v/v) formamide (200 ml) in 10 mM potassium phosphate, 10 mM EDTA (pH 7.5) (200 ml) followed by another 200 ml wash with 25% (v/v) formamide in 50 mM Tris-HCl, 0.7 M NaCl, 10 mM EDTA (pH 7.5).

1.8 Preparation of DNA-agarose Acrylamide

(i)   Dissolve DNA (~60 mg) in 0.05 M Tris-HCl, pH 7.8 (100 ml), containing acrylamide (9.7 g) and N,N'-methylenebisacrylamide (0.3 g).

(ii)   Cool the mixture to 15°C and add N,N,N',N'-tetramethylethylenediamine (TEMED) (0.8 ml).

(iii) Pour the acrylamide-DNA solution into an equal volume of 1% (w/v) agarose solution at 50°C, shake the mixture rapidly, and add ammonium persulphate (0.15 g) dissolved in a minimal volume of water with shaking.

(iv)   Allow the mixture to set for 30 min, cut into 1 cm³ cubes, and form into particles by pushing the cubes through a stainless steel mesh.

2. PROTEINS

The immobilisation of human chorionic gonadotropin (HCG) via reactive azo-dye activation has already been described (Chapter 3). Gribnau and co-workers have also immobilised other proteins, e.g., rabbit anti-HCG, sheep anti-(rabbit IgG) using the same procedure.

2.1 Immobilisation of Albumin and IgG to Agarose

The following procedures have been used to immobilise albumin and IgG.

2.1.1 *Immobilisation to Agarose via Cyanogen Bromide Activation*

(i)   Suspend agarose gel (10 g cracked wet weight) in 0.1 M NaHCO$_3$ (10 ml) and adjust the pH to 11.0 with 4.0 M NaOH.

(ii)   Adjust the temperature to 18 ± 1°C by the direct addition of crushed ice.

(iii)   Add solid CNBr (1.0 g) gradually, over the space of 2 – 3 min with continuous gentle magnetic stirring.

(iv)   Maintain the pH at 11.0 ± 0.1 by addition of 4.0 M NaOH, and keep the temperature at 18 ± 1°C. The activation takes 12 min.

(v)   Wash the activated gel alternately with 0.1 M NaHCO$_3$ and 5% (v/v) acetone (both ice cold), culminating in the former.

2.1.2 *Coupling*

(i)   Add a solution of the protein to be coupled in 10 – 20 ml of coupling buffer e.g., PBS (50 mM sodium phosphate, pH 7.4, containing 0.1 M NaCl) to the gel and mix the suspension on a Coulter rotary mixer for 24 h at 4°C.

(ii)   Wash the gel thoroughly with distilled water followed by alternate washes of 0.1 M sodium acetate, pH 4.0, containing 0.5 M NaCl, and coupling buffer, to remove ionically-bound protein.

(iii)   Block unreacted isourea groups by the addition of 0.5 M ethanolamine/HCl, pH 9.0.

2.2 Immobilisation of Proteins to Eupergit C (Oxirane Acrylic Beads)

The general procedure used is as follows.

(i)   Incubate dry Eupergit C beads (4 g) with 15 ml of a solution of protein (70 mg) in coupling buffer for 2 days at room temperature.

(ii)   After coupling, remove unbound protein by washing on a sintered funnel with distilled water, 1.0 M NaCl and further distilled water.

(iii)   Block remaining active groups by suspending the beads in a 5% (w/v) solution of mercaptoethanol, adjust to pH 8.0 with 0.5 M NaOH, for 1 day at room temperature.

(iv)   Wash the beads extensively with distilled water followed by PBS, pH 7.4.

Treatment with mercaptoethanol leaves the beads electroneutral and hydrophilic. Other reagents which can be used to block excess oxirane groups include 2-aminoethanol, glycine and thioacetic acid, all of which introduce ionisable groups into the matrix.

2.2.1 *Immobilisation of Trypsin on Eupergit C*

(i)   Dissolve trypsin (150 mg) in 1.0 M potassium phosphate buffer, pH 7.0, containing 0.05% *p*-hydroxybenzoic acid ethyl ester (4 ml).

(ii)   Add this solution to 1 g Eupergit C (w/v).

(iii)   Mix well and allow to stand without agitation at room temperature (23°C) for 72 h.

(iv) Wash thoroughly as described above.
(v) Store at 5°C suspended in phosphate buffer (0.01 M, pH 7.5), containing 0.05% (w/v) *p*-hydroxybenzoic acid ethyl ester.

2.2.2 *Activity Determination of Eupergit C-Trypsin*

The activity of Eupergit C-Trypsin can be determined using a pH-stat apparatus consisting of an autoburette operated through a pH-electrode monitoring a titration unit which allows continuous titration to a set end point (i.e., a given pH). The consumption of NaOH is simultaneously recorded *versus* time.

Substrate. Stir together casein (20 g) (E.Merck, Darmstadt, FRG), H_2O (350 ml) and 0.5 M NaOH (32 ml) at 60°C until the casein is dissolved. Cool to room temperature and adjust the pH to 8.0 by addition of 0.1 M HCl. Add water to a final volume of 500 ml. (N.B. due to aggregation of casein, this solution is slightly opaque).

Procedure. Add the substrate (20 ml) and Eupergit C-Trypsin (2 g) and stir at 37°C, pH 8.0, in the vessel provided for the pH-stat unit, continuously monitoring the degree of hydrolysis. After 10 min incubation collect the beads on a sintered glass funnel (porosity 2) and incubate for two further 10 min cycles. Repeat the incubation and filtering procedure twice more.

Activity is expressed in terms of the mean average of the activities of the second and third cycles measured. 1 Unit = μmol of peptide beads hydrolysed per min.

The caseinolytic activity of Eupergit C-Trypsin prepared as described above is about 10 U/g wet beads.

2.2.3 *Immobilisation of Pepsin on Eupergit C*

(i) Dissolve pepsin (150 mg) in 1.0 M potassium phosphate buffer (4 ml), pH 5.5 (made by mixing 91 ml 1.0 M KH_2PO_4 and 9 ml of 1.0 M K_2HPO_4).
(ii) Add to 1 g of Eupergit C.
(iii) Mix and allow to stand without agitation for 72 h at room temperature.
(iv) Wash as described above.
(v) Store in 0.1 M potassium phosphate buffer, pH 5.0, containing *p*-hydroxybenzoic acid ethyl ester (0.05% w/v).

The activity determined using haemoglobin as substrate is 2500 – 3000 U/g wet weight.

2.2.4 *Immobilisation of Lipase (from Pancreas) on Eupergit C*

(i) Dissolve lipase (200 mg) in 1.0 M potassium phosphate buffer (4 ml) (pH 7.5).
(ii) Add this solution to Eupergit C (1 g), mix well and allow to stand without further agitation at room temperature.
(iii) Wash as described above.

An average activity of 18 U/g wet beads has been reported.

2.2.5 *Immobilisation of Ribonuclease on Eupergit C*

(i)   Dissolve ribonuclease (50 mg) in 1.0 M potassium phosphate buffer, pH 7.5, (2 ml) containing *p*-hydroxybenzoic acid ethyl ester (0.05% w/v).

(ii)   Add this solution to Eupergit C (250 mg), mix well and allow to stand at room temperature (23°C) for 72 h.

(iii)   Wash as described above.

(iv)   Store at 5°C in 0.1 M potassium phosphate buffer, pH 7.5.

Determine the activity of Eupergit C-RNase in the same pH-stat-apparatus as described for Eupergit C-Trypsin.

Substrate. Dissolve RNA (10 g) from yeast (Boehringer Mannheim GmbH, FRG) in 0.2 M NaOH (125 ml). Using 0.1 M NaOH (or 0.1 M HCl), adjust the pH to 7.5, and make up to a final volume of 250 ml with distilled water.

Incubate the substrate (20 ml) with Eupergit C-RNase (500 mg) (wet beads) 37°C, pH 7.5. Titrate with 0.5 M NaOH. Using the same beads carry out three consecutive incubations. Use the linear slope of the initial reaction rate (i.e., the NaOH consumption recorded over the first 5 min) for estimation of enzyme activity; take the mean average activity of the second and third cycles. Definition of activity: 1 Unit (U) = 1 μmol of hydrolysed phosphoric acid ester bond per min.

The activity of Eupergit C-RNase obtained by the above procedure is about 60 U/g wet beads.

2.2.6 *Immobilisation of Haemoglobin on Eupergit C*

(i)   Dissolve human haemoglobin (125 mg) (Sigma, Type IV) in 1.0 M potassium phosphate buffer (pH 7.5, containing 0.1% sodium azide) (5.0 ml). Add this solution to Eupergit C (1 g); mix well and allow to stand without further agitation for 72 h at room temperature.

(ii)   Wash three times with 1.0 M NaCl and five times with distilled water.

(iii)   Mix the washed beads with mercaptoethanol (5% v/v, 2.5 ml), adjust pH to 8.0 and allow to stand overnight at room temperature.

(iv)   Wash with distilled water (10 times).

(v)   Pack the beads into a small column and wash as follows: 0.1 M potassium phosphate buffer, pH 7.5 (50 ml); 0.5 M potassium phosphate buffer, pH 7.5 (50 ml); 3.5 M sodium thiocyanate (50 ml); and finally PBS (0.01 M sodium phosphate, pH 7.2, containing 0.15 M NaCl), until the absorbance at 418 nm of the eluate is less than 0.01.

2.2.7 *Affinity Purification of Anti-human Haemoglobin Antibodies from Goat Serum using Eupergit C-immobilised Haemoglobin*

(i)   Carefully apply antiserum (goat anti-human haemoglobin) (1 ml) on top of the beads prepared as described above.

(ii)   Apply PBS, pH 7.2 (1 ml).

(iii)   Shake the column (or rotate closed column) for 1 h.

(iv)   Wash with PBS until the absorbance at 280 nm is less than 0.01.

(v)   Elute with 3.5 M sodium thiocyanate (30 ml) at a flow-rate of 15 – 30 ml/h.

(vi) Collect the protein-containing fractions and dialyse against PBS until a test for thiocyanate is negative (Test: 1 ml eluate, 0.1 ml 1.0 M HCl and 0.05 ml 4.5% $FeCl_3$ in 0.02 M HCl).

(vii) Concentrate by ultrafiltration or vacuum dialysis. Starting from a volume of 30 – 50 ml, vacuum dialysis takes over 24 h to reach a final volume of 1 ml; nevertheless, this method allows high yields of pure antibody.

(viii) Store the product at −20°C.

(ix) Characterise for purity and antibody content by one of the following methods:

    (a) Ouchterlony's immunodiffusion method

    (b) Immunoelectrophoresis according to Grabar-Williams

    (c) Flat bed electrophoresis (isoelectric focusing).

2.3 Carrier-bound α_2-Macroglobulin

α_2-Macroglobulin forms high affinity bonds with Zn^{2+}, so that the protein can be bound by means of a zinc chelate to an agarose carrier (3). Using carrier-bound α_2-macroglobulin it is possible to remove almost all known endoproteinases, including carboxyl-, thiol-, serine- and metal proteinases, from solutions. The adsorbent is suitable for the removal of most interfering proteinases, e.g., from crude extracts, enzyme solutions or cell culture media in one-step process.

2.4 Immobilisation of Urease on to Nylon

(i) Partially hydrolyse a 2 m length of nylon tubing made from 'Type 6' nylon (0.1 cm internal diameter) on its inside surface by perfusion for 60 min at 30°C with 3.0 M HCl at a flow-rate of 2 ml/min.

(ii) Stop the hydrolysis by washing through the tube with water, followed by a 2.5% (v/v) solution of glutaraldehyde in bicarbonate buffer, pH 9.4, at 0°C at a flow-rate of 2 ml/min.

(iii) Wash the tube through with phosphate buffer, pH 8.0 at 0°C, containing 1 mM EDTA and then perfuse for 30 min at 0°C with a 7.5% (w/v) solution of urease in the phosphate buffer containing EDTA (1 mM) and mercaptoethanol (10 mM).

(iv) Wash the inside surface of the tube free of any adherent enzyme by perfusion in turn with 0.1 M $NaHCO_3$, 1.0 M NaCl and water.

(v) Store the tube at 4°C in water when not in use.

This method yields approximately 62.5 μg protein bound per metre length of tubing (4).

2.5 Immobilisation of Pepsin

2.5.1 Immobilisation to Spheron-type Gels

After activation with cyanogen bromide bind either 1,6-diaminohexane or ϵ-aminocaproic acid to hydroxyalkyl methacrylate gels of the Spheron type by the procedure of Cuatrecasas (5). The derivatives thus obtained are denoted NH$_2$-Spheron and COOH-Spheron.

(i)   Suspend 4.8 ml of NH$_2$-Spheron or COOH-Spheron in 15 ml of 0.1 M sodium acetate, pH 4.0, containing 300 mg of pepsin.

(ii)   After stirring for 5 min, add 100 mg N-ethyl-N'-(3-dimethylaminopropyl)-carbodiimide hydrochloride to each suspension and continue stirring with a magnetic stirrer at 4°C.

(iii)   After 22 h decant both samples; wash several times with 0.1 M sodium acetate, pH 4.1, containing 1.0 M sodium chloride, and wash on the column alternately with the above buffer and 3.0 M urea, pH 3.0.

(iv)   Finally wash the gels with 0.01 M sodium acetate, pH 4.1, and with 0.1 M acetic acid and store in the wet cake form at 4°C.

The amount of fixed enzyme may be determined from that of acid and neutral amino acids present in the hydrolysate after acid hydrolysis of dried gels.

2.5.2 *Preparation of Pepsin Coupled to Glass*

Silanise porous 96% silica glass particles, 550 μm average pore diameter, 40/60 mesh particle size, with γ-aminopropyltriethoxysilane as described by Weetall (6). Wash the alkylaminosilane-glass derivative with acetone and air dry.

Couple the pepsin to the aminofunctional glass as follows:

(i)   Suspend 1 g of glass in 10 ml water containing 50 mg pepsin.

(ii)   Add 40 mg of l-cyclohexyl-3-(2-morpholinoethyl)carbodiimide, methyl-*p*-toluenesulphonate and adjust the pH to and maintain at 4.0 with HCl for 30 min at room temperature.

(iii)   Continue the reaction at 6°C overnight, then harvest the solid by filtration, wash exhaustively with 0.01 M HCl, and store as a moist filter cake at 4°C.

2.5.3 *Immobilisation of Pepsin to Agarose*

(i) *Preparation of the carrier.* Activate cross-linked agarose (200 mg dry weight) with CNBr and allow to react with 100 mg *p*-phenylenediamine or 25 mg 4,4'-methylene dianiline in 10 ml 0.1 M NaHCO$_3$. Bubble nitrogen through the solution for 15 min and allow the coupling to proceed in a closed vessel for 16 h in the dark at room temperature with slow rotation. Wash the polymer in a column with the following buffers: 0.1 M sodium borate buffer (1.0 M in NaCl), pH 8.5 (24 h); 0.1 M sodium acetate buffer (1.0 M in NaCl), pH 4.5 (24 h); 0.01 M sodium acetate buffer, pH 4.5 (24 h) and finally, distilled water (4 h). Store the polymer in 0.01 M acetate buffer, containing NaN$_3$ (0.02% w/v).

(ii) *Coupling of pepsin.* Suspend the activated agarose (50 mg) in 4 ml distilled water in a small pH-stat vessel; add 25 μl acetaldehyde, followed by 10 – 25 mg pepsin and 25 μl isocyanide. Keep the pH at 5.8 by means of a pH-stat (Radiometer, Copenhagen) for 6 h. Wash the product on a G3 glass filter with water and then in a small column with the following buffers: 0.1 M sodium citrate buffer (1.0 M in NaCl), pH 4.0 (48 h); 0.1 M sodium   acetate buffer

(1.0 M in NaCl), pH 5.4 (24 h); and finally 0.01 M sodium citrate buffer, pH 4.0. Store the conjugates at 4°C in 0.01 M sodium citrate, pH 4.0, containing sodium azide (0.02% w/v).

2.6 Immobilisation of Aspartate Amino Transferase

Mix CNBr-activated Sepharose 4B (0.5 g wet weight) with 1.0 mg of aspartate amino transferase dissolved in 0.1 ml of 0.05 M Tris-HCl buffer, pH 7.0. Allow the coupling reaction to proceed for 16 h at 4°C. Wash the immobilised enzyme thoroughly with 0.1 M potassium phosphate buffer of pH 8.5 and 5.5 alternately.

2.7 Immobilisation of Tryptophanase

Mix CNBr-activated Sepharose (10 ml) with 40 mg of tryptophanase dissolved in 0.1 M potassium phosphate buffer, pH 7.0 (5.0 ml). Gently agitate the mixture for 24 h at 4°C. By this treatment, about 80% of the initial amount of protein is immobilised on Sepharose. The immobilised enzyme preparation, when used in continuous enzyme reactions over 5 days at 37°C, shows only a slight decrease ($\sim 3\%$) in enzymatic activity over this period.

2.8 Coupling of Rabbit IgG to Hydrazide Beads

Hydrazide derivatised polystyrene beads are available from Pierce Chemical Co.

(i)   To 4 g of hydrazide beads (~ 25) add 5 ml of 12.5% glutaraldehyde solution (2.5 ml of 25% glutaraldehyde diluted to 5 ml with 0.1 M sodium phosphate, pH 7.0).

(ii)   Shake very gently for 2 h at room temperature.

(iii)   Wash with 100 ml of water in a Buchner funnel (without filter paper) followed by 0.1 M sodium phosphate, pH 6.0 (20 ml).

(iv)   Add the glutaraldehyde-activated beads to a solution of rabbit IgG (2.5 mg dissolved in 5.0 ml of 0.1 M sodium phosphate, pH 6.0).

(v)   Follow with the addition of 1 mg of sodium cyanoborohydride. (This step reduces the Schiff's bases formed above to stable secondary amine linkages.)

(vi)   Shake the reaction mixture gently at room temperature overnight.

(vii)   Wash the IgG-coupled polystyrene beads with 0.1 M sodium phosphate, pH 6.0 (50 ml), followed by 0.1 M sodium bicarbonate (100 ml).

(viii) Add the IgG-coupled polystyrene beads to 0.1 M sodium bicarbonate (5.0 ml) containing sodium borohydride (~ 1 mg).

(ix)   Shake gently for 15 min. (This step reduces residual aldehydes not coupled to protein.)

(x)   Wash the IgG-coupled beads with 0.1 M sodium carbonate (100 ml) followed by water (100 ml).

(xi)   Dry the IgG-coupled beads with a paper towel and store dry at 4°C.

2.9 Use of BSA-based Silica Columns for Optical Resolution by h.p.l.c.

by S.Allenmark and B.Bomgren

This section describes the experimental use of BSA-silica columns in h.p.l.c. for analytical scale optical resolutions. The technique of column packing has also been included for those who have access to the necessary equipment and want to make their own columns. After a general outline of the method and its applicability, specific experimental conditions are given for a number of optical resolutions selected as examples of different types of compounds. Finally, a list is given of the compounds resolved by the technique to date.

2.9.1 *Introduction*

Serum albumins are generally regarded as transport proteins in the body and their binding properties, i.e., the 'affinity' for various endogenous compounds and drugs, have provided an attractive area of research among pharmacologists and biochemists. Equilibrium studies in solution have demonstrated enantioselective properties of albumins (7), but the chromatographic use of these phenomena has not been investigated until quite recently (8 – 13). The technique of using a protein such as bovine serum albumin (BSA) as a stationary phase in liquid chromatography can be regarded as a 'reversed' affinity chromatography, because here the protein is the immobilised component and not the small ligand counterpart. Starting with BSA coupled to agarose in plastic or glass columns and using a peristaltic pump for mobile phase delivery, it has been possible to synthesise a chromatographic material for h.p.l.c. based on a 10 μm silica as the support (14) (Resolvosil 10 μ-BSA, now available from Macherey-Nagel GmbH, Düren, FRG).

2.9.2 *Principle*

Retention, and in many cases differential retention between two optical isomers, can be regulated very efficiently by the mobile phase, which consists of a buffer system of a pH between 4 and 9. The enantioselective properties of the stationary phase may be caused by a combination of Coulombic and hydrophobic interactions in the binding of solute; the net effect upon the antipodes being different for steric reasons. Binding to the stationary phase, and hence the observed retention

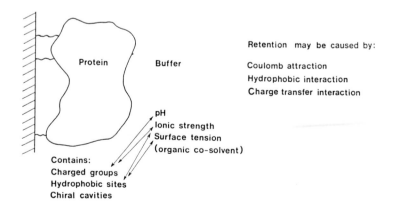

Figure 1. Factors influencing retention on BSA-based silica columns.

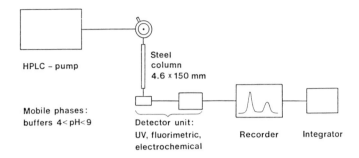

Figure 2. H.p.l.c. system for optical resolution using BSA-silica columns.

on the column, can be greatly influenced by changes in pH, ionic strength and organic co-solvent modifier of the mobile phase (*Figure 1*). The columns have very limited capacity and can only, at present, be used for analytical scale separations; the amount of solute injected will typically be of the order of $0.5 - 5$ nmol.

2.9.3 *Instrumentation*

The components needed are those commonly found in the basic parts of a commercially available h.p.l.c. system. The high-pressure pump can be of the relatively inexpensive single piston type, but is preferably combined with an external pulse-dampening device. An injection valve provided with a 10 or 20 μl loop and connecting the solvent delivery system to the (steel) column packed with BSA-silica will be necessary. The column outlet can be coupled to various types of detectors (u.v.-visible, fluorimetric or electrochemical). By far the most versatile detector is a variable wavelength u.v. monitor, although for certain applications requiring very high sensitivity an electrochemical or fluorimetric detector might be a better choice. A simple single-channel potentiometric X-Y recorder is the last essential component; however, an electronic integrator is valuable for the determination of peak area ratios when determining enantiomer composition or optical purity of injected samples. The total chromatographic system is shown in *Figure 2*. So far, all optical resolutions have been performed by means of isocratic elution and therefore there is no need for any gradient making accessories in the system.

2.9.4 *Column Packing*

A packing pump, driven by compressed air and connected to a stainless steel reservoir of about 75 ml volume, will provide adequate conditions for slurry-packing (15) of the analytical columns.

(i) First clean all parts of an empty analytical column carefully, preferably by ultrasonication in methanol or acetone.

(ii) After drying, assemble the column except for the inlet end-cap. The inner dimensions of the column tube should be $4 - 5$ mm x $150 - 200$ mm.

(iii) For a 4.6 x 150 mm column measure out BSA-silica slurry corresponding to approximately 2 g of dry material volumetrically and disperse the silica (in a graduated cylinder) in 50 mM phosphate buffer, pH 7.0 containing 1% (v/v) 1-propanol (75 ml).

(iv) First, let this slurry stand for about 20 min and discard the supernatant. Replace the liquid with fresh buffer if noticeable turbidity exists (due to fine particles). This procedure may be repeated in order to avoid clogging of the outlet of the column.

(v) Then fill the slurry reservoir and connect it to the inlet port of the column (the column should be upside down).

(vi) Connect suitable tubing to the column outlet for solvent collection.

(vii) With the pump ready to deliver the same phosphate buffer as described above, start the pump and pack the column at about 5000 p.s.i. (340 bar) until approximately 250 ml of the liquid has passed through the column.

(viii) After inverting the column by 180° (inlet upwards) the pump is stopped and subsequent to pressure equilibration (wait for 5 min) the column is removed from the reservoir and sealed with the inlet port. It can then be placed directly into the chromatographic system for equilibration with the mobile phase.

2.9.5 *Experimental Procedures for Optical Resolution*

(i) *Resolution tests and optimisation procedures.* Although h.p.l.c. on BSA-silica columns will never serve as a universal technique for optical resolution of racemic organic compounds, a surprising number of compounds of different types have been fully or partially resolved into enantiomers by this method. At present, we know little about the criteria that have to be fulfilled by the molecular structure of the racemic compound to promote an enantiomeric selection by the BSA. From our experience to date, compounds containing aromatic as well as polar substituents (such as amide, ester, keto, carboxyl or sulphoxide functions) are good candidates. Strong retention on the column does not ensure optical resolution [which is often caused by one dominating binding interaction (16)]; some fast-eluting compounds show complete optical resolution, whereas many strongly-retained compounds are not optically resolved at all.

To test whether a particular racemic compound can be resolved, it is best to follow the procedure in *Figure 3*. Concentrations of propan-1-ol higher than 7% should be avoided and are normally unnecessary except for very strongly hydrophobic compounds. If no suggestion of even partial resolution is found under any of the conditions outlined in *Figure 3*, proceed no further. If a partial resolution is observed, however, an improvement can usually be obtained either by a further change of pH, or by a change of the ionic strength, or by a combination of the two.

(ii) *Procedure for optical resolutions.* Equilibrate the column with the appropriate mobile phase. Because the column responds rather slowly to a mobile phase change, completely reproducible retention parameters will not be obtained unless there has been sufficient time for re-equilibration. About 1 h will be sufficient in most instances. A flow-rate of 1.0 – 1.5 ml/min is appropriate during equilibration and chromatography. The mobile phase may be recirculated in the

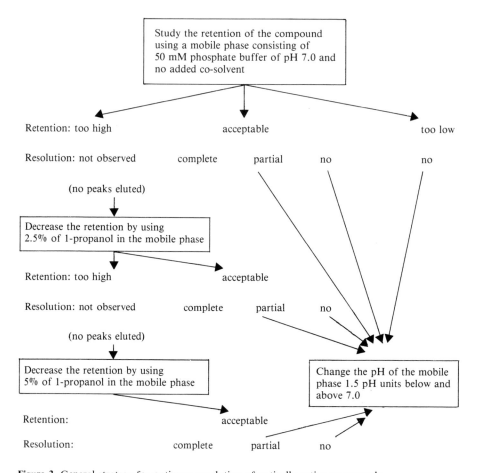

Figure 3. General strategy for optimum resolution of optically active compounds.

system. Select a wavelength suitable for the compound of interest. Prepare an approximately 100 μM solution of the racemic compound in a water-miscible solvent (preferably the mobile phase, but ethanol, acetone, dioxane or similar solvents can also be used). Attenuate the integrator and start the recorder (speed 5 mm/min). Inject the sample onto the column *via* the sample loop. Simultaneously, mark the chart paper and start the integrator.

2.9.6 *Examples*

(i) *N-Aroyl-D,L-amino acids.* Amino acids may be converted to their N-aroyl derivatives by reaction with the appropriate aroyl chloride in a solution of sodium carbonate. Many of these derivatives are readily optically resolved (14). Generally, N-naphthoyl derivatives are not eluted from the column within a reasonable time unless about 5% (v/v) 1-propanol is added to the mobile phase. Retention times increase with decreasing pH.

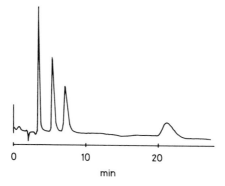

Figure 4. Resolution of N-benzoyl-D,L-serine + N-benzoyl-D,L-alanine (mixture). Conditions used: 50 mM phosphate buffer, pH 5.7, 1.0 ml/min, u.v. 225 nm, sample 40 μM (each component), 20 μl injected.

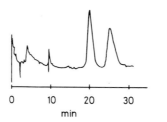

Figure 5. Optical resolution of N-(2,4-dinitrophenyl)-D,L-serine. Conditions used: 50 mM phosphate buffer, pH 5.7, 1.0 ml/min, u.v. 254 nm, sample 66 μM, 10 μl injected.

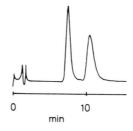

Figure 6. Optical resolution of N-acetyl-D,L-tryptophan ethyl ester. Conditions used: 50 mM borate buffer, pH 8.25, 1.0 ml/min, u.v. 225 nm, sample 75 μM, 10 μl injected.

Figure 4 shows optical resolution of a mixture of N-benzoyl-D,L-serine + N-benzoyl-D,L-alanine.

(ii) *N-Aryl-D,L-amino acids.* Examples belonging to this group are N-(2,4-dinitrophenyl)-D,L-amino acids, which are easily prepared from 2,4-dinitrofluorobenzene and the corresponding amino acid. Optical resolution of N-(2,4-dinitrophenyl)-D,L-serine is shown in *Figure 5*.

(iii) *Tryptophan derivatives.* Among these, BSA-silica columns have been shown to resolve, in addition to the parent compound (D,L-tryptophan), 5-hydroxy-D,L-tryptophan, N-formyl- and N-acetyl-D,L-tryptophan as well as N-acetyl-D,L-tryptophan ethyl ester (17). A complete optical resolution of the ester is shown in *Figure 6*.

113

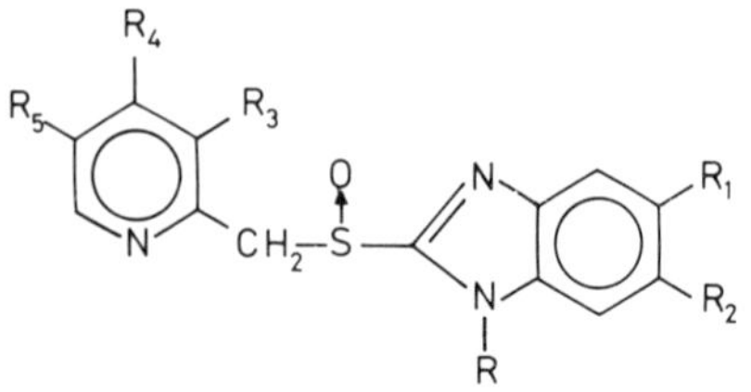

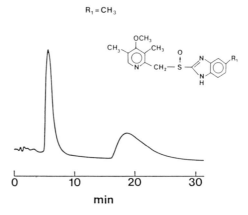

Figure 7. Structure of resolvable aromatic sulphoxide.

$R_1 = CH_3$

Figure 8. Optical resolution of Ia ($R_1 = R_3 = R_5 = CH_3$, $R = R_2 = H$, $R_4 = CH_3O$) (*Figure 7*). Conditions used: 80 mM phosphate buffer, pH 5.80, 2.0 ml/min, u.v. 225 nm, sample 500 μM, 10 μl injected.

(iv) *Aromatic sulphoxides*. Sulphoxides having unequal S-substituents are intrinsically chiral and therefore possible to resolve into optical antipodes (*Figures 7 and 8*). Compounds of the general type I (R = H) were shown to be resolvable on BSA-silica columns (13,18).

Table 1 shows the range of compounds successfully separated to date.

3. IMMOBILISATION OF LECTINS

3.1 Immobilisation of Wheat Germ Lectins to Sepharose

Incubate 1 g (dry weight) of CNBr-activated Sepharose (as described in Chapter 3) at 4°C with gentle stirring for 24 h with 25 mg of wheat germ agglutinin (WGA) in 0.1 M NaHCO$_3$ containing 0.5 M NaCl and chitin oligosaccharides at a concentration of 10–50 mM (based on reducing equivalents using glucose as a standard). An approximate quantitation of the bound lectin may be obtained by measuring the A_{280} of the protein not bound to the Sepharose. Approximately 90–95% of the wheat germ agglutinin should be bound to the support. Store the WGA-Sepharose at 4°C in PBS containing sodium azide (0.5% w/v) and N-acetyl-D-glucosamine (0.1 M).

Table 1. List of Racemic Compounds Fully or Partially Separated into Enantiomers on BSA-silica Columns.

1. *Amino acids*
   tryptophan
   5-hydroxytryptophan
   kynurenine
   3-hydroxykynurenine

2. *Amino acid derivatives*
   N-formyltryptophan
   N-acetyltryptophan
   N-acetyltryptophan ethyl ester
   N-benzoylserine
   N-(4-nitrobenzoyl)serine
   N-(2,4-dinitrophenyl)serine
   N-benzoylalanine
   N-(4-nitrobenzoyl)alanine
   N-(2,4-dinitrophenyl)alanine
   N-(2-naphthoyl)alanine
   N-(2-thenoyl)alanine
   N-benzoylphenylalanine
   N-(2-naphthoyl)phenylalanine
   N-benzoylthreonine
   N-(2,4-dinitrophenyl)threonine
   N-benzoylmethionine
   N-benzoylleucine
   N-(2,4-dinitrophenyl)aspartic acid
   N-(2,4-dinitrophenyl)glutamic acid

3. *Coumarin derivatives*
   warfarin
   phenprocoumon

4. *Sulphoxides*
   Compounds of type I:
   (a) $R = H$, $R_3 = R_5 = CH_3$, $R_4 = CH_3O$: $R_1 = CH_3$, $R_2 = H$; $R_1 = CH_3O$, $R_2 = H$; $R_1 = R_2 = H$; $R_1 = Cl$, $R_2 = H$; $R_1 = Br$, $R_2 = H$; R_1 $(CH_3)_2CH$, $R_2 = H$; $R_1 = (CH_3)_3C$, $R_2 = H$; $R_1 = CH_3$, $R_2 = CH_3OCO$;
   (b) $R = H$, $R_3 = R_5 = H$, $R_4 = CH_3$: $R_1 = CH_3$, $R_2 = CH_3OCO$; 2-methylsulphinylbenzoic acid.

3.1.1 *Coupling to Agarose*

(i)   Wash CNBr-activated agarose beads on a Buchner funnel with $5-6$ volumes of ice-cold distilled water and then with 0.5 M $NaHCO_3$, pH 8.5.

(ii) Add the lectin solution in 0.5 M $NaHCO_3$.

(iii) Stir, or preferably rock the slurry gently overnight at room temperature. This yields a gel containing $1-5$ mg lectin per gram of moist weight gel.

(iv) Block unreacted iminocarbonate by similar treatment with a saturated solution of glycine for 24 h. The potentially ionogenic carboxyl group of glycine is screened by the basic imino group in the imidocarbonate.

(v)   Wash the resulting immobilised lectin adsorbent thoroughly and equilibrate with $20-50$ mM phosphate buffer (pH 7.2) containing 0.15 M sodium chloride.

(vi) Apply the sample to be chromatographed and wash the column with the equilibrating buffer until the absorbance at 280 nm returns to the baseline value.

(vii) Elute the bound material with the same buffer containing 0.1 – 0.3 M of an appropriate sugar.

3.1.2 *Immobilisation of Plant Lectin Adsorbents to Agarose*

Activate 40 ml of agarose with CNBr (6 g) and allow to react with 60 ml of ice-cold 0.1 M sodium phosphate, pH 7.4, containing concanavalin A (500 mg) and α-methyl-D-mannopyranoside (0.1 M). After 16 h at 4°C, add glycine (2 g) and continue the incubation for 8 h at room temperature. This adsorbent contains 5.5 mg of protein/ml of gel. Wheat germ agglutinin (1.4 mg/ml) is similarly coupled to activated agarose in the presence of 0.1 M N-acetyl-D-glucosamine; 1.1 mg of protein is coupled per ml of gel.

3.2 **Immobilisation of Heparin**

3.2.1 *Immobilisation of Heparin on Eupergit C*

(i) Dissolve heparin (Sigma, grade 1, from porcine intestinal mucosa) (100 mg) in water (6 ml) and adjust the pH to 10 – 11 by addition of 1.0 M NaOH.

(ii) Bring the total volume to 8 ml and add this solution to Eupergit C (2.0 g).

(iii) Allow the mixture to stand without further agitation for 24 h at room temperature, then wash with water until the pH is neutral.

(iv) If necessary, remove residual oxirane groups by treatment with mercaptoethanol (see Chapter 1).

3.2.2 *Heparin-Ultrogel*

Heparin-Ultrogel A4R combines the properties of conventional soluble heparin with those of Ultrogel A4 (4% agarose beads) and is available from Reactifs IBF. The heparin used is prepared from porcine intestinal mucosa and is purified so as to retain full biological activity. It has a molecular weight of about 10 000 and an activity greater than 150 USP units/mg. The exclusion limit of the Ultrogel A4 matrix is several millions, thus making Heparin Ultrogel A4R suitable for the purification of very high molecular weight compounds, such as serum lipoproteins.

Heparin is immobilised with the coupling agent EEDQ on a 6-carbon spacer arm. The latter is fixed to the support after activation and cross-linking of the agarose beads with epichlorohydrin. The resulting complex is extremely stable, even in the presence of strong dissociating agents. 4 – 6 mg of heparin are immobilised per ml of gel, with an activity of 600 – 900 USP units/ml.

The use of heparin-Sepharose in affinity chromatography is further described in Section 5.

4. PURIFICATION OF γ-GLUTAMYLTRANSFERASE BY IMMUNO-AFFINITY CHROMATOGRAPHY

by A.Szewczuk and E.Prusak

4.1 Introduction

γ-Glutamyltransferase (γ-glutamyltranspeptidase, GGT, EC 2.3.2.2) catalyses the transfer of the γ-glutamyl group from a variety of donors, including glutathione, to a wide range of amino acid and peptide acceptors. It also catalyses the hydrolysis of γ-glutamyl compounds. GGT is a heterodimeric membrane glycoprotein which can be solubilised either with detergents yielding the 'heavy' (detergent-solubilised) form or by proteolytic digestion yielding the 'light' (protease-solubilised) form of the enzyme (19,20).

Many methods have been reported for the purification of the various forms of GGT from different tissues and body fluids; in general, these require multiple steps and recoveries are relatively low. Furthermore, GGT has multi-molecular forms with different isoelectric points, due to variations in carbohydrate and sialic acid content, which are not effectively separated by ion-exchange chromatography or by affinity chromatography on immobilised Concanavalin A. Therefore, the purification of GGT forms by immunoaffinity chromatography seems to be an effective alternative method.

4.2 Basic Techniques

(i) *Assay of GGT activity*. Determine the activity of γ-glutamyltransferase spectrophotometrically with γ-L-glutamyl-*p*-nitroanilide as substrate and glycylglycine as acceptor (21). The substrate solution contains: 5 mM γ-L-glutamyl-*p*-nitroanilide (13.3 mg in 10 ml), and 50 mM glycylglycine (66 mg in 10 ml) in 0.1 M Tris-HCl buffer, pH 7.6. To 0.4 ml of the substrate solution add 0.1 ml of diluted GGT preparation (0.005 – 0.2 U) and incubate the mixture at 37°C for 1 – 10 min. Stop the enzymatic reaction with 1.0 M acetic acid (2 ml) and measure the optical density at 407 nm against a blank sample prepared in the same manner but with GGT added after the acetic acid. Prepare a standard curve for *p*-nitroaniline (0 – 0.2 μmol/sample). Express the enzyme activity in units (U) equal to the number of μmol of *p*-nitroaniline liberated per min.

(ii) *Protein concentration*. Measure protein concentration colorimetrically with Folin reagent (22), using crystalline bovine serum albumin as a standard. If the protein sample contains Triton X-100 perform the assay with sodium dodecylsulphate (23).

(iii) *Gel electrophoresis*. Perform polyacrylamide gel electrophoresis in 7% (w/v) gels as described by Davis (24): stain the gels either for protein (24) or for GGT activity by a fluorescence method using 7-(γ-L-glutamyl)-4-methylcoumarylamide as substrate (25). Prepare the fluorogenic substrate by dissolving 7-(γ-L-glutamyl)-4-methylcoumarylamide (1.2 mg) in 2.0 M NaOH (0.02 ml); dilute with 50 mM glycylglycine in 0.1 M Tris-HCl buffer, pH 7.8 (10 ml). Incubate the gels with the fluorogenic substrate; 7-amino-4-methylcoumarin, liberated by GGT, is visible under the u.v. lamp as a bright blue band.

(iv) *Double immunodiffusion*. Use 1.2% (w/v) agar gels for double immuno-diffusion (26). After formation of precipitation arcs, wash the gels with PBS (0.15 M sodium chloride in 10 mM phosphate buffer, pH 7.4) for 3 days and then stain as described for polyacrylamide gels.

(v) *Antigens*. Antigens used for immunisation should be highly purified. They can be obtained either by immunoaffinity chromatography or by 'classical methods'. Pure GGT from bovine kidney can be isolated by solubilisation of the enzyme with sodium deoxycholate, followed by treatment with n-butanol, frac-tionation with ammonium sulphate and chromatography on DEAE-cellulose (27). The GGT preparation obtained after separation on Sepharose 6B is sub-jected to proteolytic digestion with bromelain and the resulting GGT (light form) purified by polyacrylamide gel electrophoresis (28). Pure GGT from rat kidney-enzyme III can be obtained (29) by a similar procedure.

4.3 Production and Isolation of Antibodies

Polyclonal antibodies directed against bovine kidney GGT or rat kidney GGT (light forms) can be produced in rabbits (females, 6−8 weeks of age) which are immunised by injection into the hind foot pads with an emulsion consisting of 50 μg of enzyme in 0.5 ml PBS with 0.5 ml of Freund's complete adjuvant per animal. After 2 weeks inject a similar amount of antigen (50 μg GGT in 0.5 ml PBS), emulsified well with 0.5 ml of Freund's incomplete adjuvant, subcutan-eously in multiple sites into the neck. It is advantageous to repeat the last injec-tion after 2 weeks. Collect the rabbit blood after 10−15 days; incubate at 37°C for 2 h, then leave at 4°C overnight and separate antiserum by centrifugation at 3000 g for 30 min. In double immunodiffusion tests, the antiserum at 1:128 dilu-tion (with PBS) still gives precipitin arcs with diluted GGT (0.025 mg/ml).

The immunoglobulin G (IgG) fraction can be isolated from antiserum by the procedure described below (30).

(i)   Adjust the serum to 50% saturation with ammonium sulphate (31.3 g per 100 ml).

(ii)   After 1 h collect the precipitated protein by centrifugation (10 000 g for 60 min) and dissolve in a small volume of 10 mM potassium phosphate, pH 6.8.

(iii)   Adjust the volume of the solution to that of the original antiserum sample prior to salt addition and re-precipitate the immunoglobulin fraction by the addition of solid ammonium sulphate to 50% saturation.

(iv)   Dissolve the precipitate as before and dialyse the solution against three changes of 100 volumes each of 10 mM potassium phosphate, pH 6.8.

(v)   For further purification pass the sample through a DEAE-cellulose column (DE-52, Whatman) previously equilibrated with 10 mM potassium phosphate pH 6.8, using 5−8 ml of the packed cellulose per ml of anti-serum to be fractionated.

(vi)   Wash the column with the equilibration buffer and collect fractions.

IgG is not retarded on the cellulose column; identify IgG-containing fractions by measuring absorbances at 280 nm, taking the absorbance of a 0.1% solution in a 1 cm light path to be 1.45. Combine fractions with an IgG concentration higher than 0.5 mg/ml and concentrate at least 30-fold by ultrafiltration. Antibody preparations thus obtained can be stored at $-20°C$ without significant loss in immunological activity for several months. Using this procedure it is possible to obtain $400-600$ mg IgG from one rabbit.

4.4 Preparation of Immunoadsorbents

Immunoadsorbent can be obtained by immobilisation of the above antibody preparation on CNBr-activated Sepharose (31). The standard procedure is as follows.

(i)   Wash 10 g of a slurry of Sepharose 4B (Pharmacia, Uppsala, Sweden) three times with distilled water on a filter-funnel, then place in a beaker and suspend in 2.0 M sodium carbonate (10 ml) by mixing with a magnetic stirrer under a ventilated hood.

(ii)   Immediately add cyanogen bromide solution (1 g in 1 ml of acetonitrile) and stir vigorously for 2 min at room temperature.

(iii) Transfer the suspension to a filter-funnel and wash rapidly under suction with ice-cold deionised water (100 ml) and then with 0.25 M carbonate buffer, pH 9.5 (100 ml).

(iv)   Quickly transfer the gel to a flask ($50-75$ ml) containing 150 mg of antibody in 0.25 M carbonate buffer, pH 9.5 (10 ml).

(v)   Shake the flask gently to keep the gel in suspension at $4°C$ for 16 h.

(vi)   Filter the gel, and wash with the carbonate buffer containing 0.5 M sodium chloride; to destroy the remaining active groups leave the gel in Tris-HCl buffer, pH 8.6, (50 ml) for 2 h at room temperature.

(vii) Wash the gel again with 0.25 M carbonate buffer containing 0.5 M sodium chloride and then with PBS containing sodium azide (0.02% w/v). The gel can be stored at $4°C$ for several weeks without significant decrease in the activity of immobilised antibody.

4.5 Purification of GGT (Light Form) from Animal Tissues

The method is based on bromelain digestion of the 100 000 g pellet, isolated from tissue homogenate; the soluble enzyme thus obtained is further purified by affinity chromatography on a specific immunoadsorbent column.

4.5.1 *Soluble Preparation of GGT (Light Form)*

Fresh tissues (kidneys, livers, hepatomas) from rats and mice are obtained after decapitation of animals. Bovine tissues are obtained from the slaughter-house. Tissues can be stored below $-20°C$ for several months without loss of GGT activity.

(i)   After thawing, mince tissues and homogenise with cold PBS in a Waring Blendor using 4 volumes of PBS per volume of tissue.

(ii)   Centrifuge the liquid pulp at 3000 g for 15 min and discard sediment, then centrifuge the supernatant at 100 000 g for 60 min in a cooled ultracentrifuge.

(iii)   Suspend the separated pellet in 50 mM phosphate buffer, pH 7.4, containing 1 mM EDTA ($\sim$ 5 vols).

(iv)   Adjust the protein concentration of the suspension to about 20 mg/ml with the phosphate buffer-EDTA and add 2-mercaptoethanol to a final concentration of 10 mM.

(v)   Add bromelain (Sigma, specific activity 2310 U/mg) (1 mg/10 mg of protein) and incubate the suspension at 37°C for:

   3 h for solubilisation of kidney and hepatoma GGT
   15 h for solubilisation of bovine liver GGT.

(vi)   Centrifuge the resulting suspension at 100 000 g for 60 min. The supernatant, which contains solubilised GGT, may be stored at -30°C before further purification.

4.5.2   *Immunoaffinity Chromatography*

(i)   Pack a column with a suspension of immunoadsorbent and wash with several volumes of PBS. Use 5 ml (1.2 x 4.5 cm) or 1.5 ml (0.8 x 3 cm) columns of packed immunoadsorbents.

(ii)   To the solubilised GGT preparation add sodium chloride (85 mg/10 ml), pass the solution through the immunoadsorbent column at room temperature and collect fractions (5 ml or 1 ml with flow-rates of 40 or 10 ml/h, for 5 ml and 1.5 ml columns, respectively).

(iii)   Apply approximately 1000 U of rat kidney GGT per ml of immobilised anti-rat kidney GGT antibody and 400 U of bovine kidney GGT per ml of immobilised anti-bovine kidney GGT antibody. Fractions with GGT not retarded on immunoadsorbent can be re-chromatographed again.

(iv)   Wash the column with PBS (10 vols) and elute GGT either with 0.1 M ethanolamine (flow-rate 10 and 3 ml/h) or with deionised water (flow-rate $\sim$ 2 and 0.5 ml/h).

(v)   Ethanolamine eluted fractions should be neutralised with 0.5 M acetic acid in the presence of phenolphthalein (one drop of 0.1% w/v solution in ethanol): measure GGT activity in small samples (1 – 10 μl).

(vi)   Combine fractions with activity exceeding 2 U/ml and either purify by re-chromatography or concentrate by ultrafiltration.

4.5.3 *Re-chromatography*

To the GGT preparation eluted from immunoadsorbents, add sodium chloride (85 mg/10 ml), and re-chromatograph the material under the same conditions as for chromatography. Re-chromatography increases the specific activity of the purified GGT as is demonstrated for bovine kidney GGT and mouse kidney GGT (*Tables 2* and *7*).

4.5.4 *Regeneration of Immunoadsorbents*

Immunoadsorbent columns after elution of GGT can be regenerated by washing with 0.2 M of glycine/HCl buffer, pH 2.2 (10 volumes) and then with PBS. Thus

Table 2. Purification of GGT (Light Form) from 400 g of Bovine Kidney.

Preparation	Volume (ml)	Protein (mg)	Total activity (U)	Specific activity (U/mg)	Purification (-fold)	Yield (%)
Homogenate	1300	31 200	10 920	0.35	1	100
100 000 g pellet suspension	360	7380	10 260	1.4	4	94
After solubilisation with bromelain	275	1790	7840	4.4	12.6	72
Eluate from immunoadsorbent[a]	33	6.9	5915	850	2430	54
After re-chromatography[a]	24	4.0	5030	1230	3515	46

[a]Immobilised anti-bovine kidney GGT antibody (three 5 ml columns).
Elution of GGT with 0.1 M ethanolamine.

Table 3. Purification of GGT (Heavy Form) from 80 g of Bovine Kidney.

Preparation	Volume (ml)	Protein (mg)	Total activity (U)	Specific activity (U/mg)	Purification (-fold)	Yield (%)
Homogenate	260	6240	2180	0.35	1	100
100 000 g pellet suspension	72	1480	2050	1.4	4	94
After solubilisation with Triton X-100	60	710	1350	1.9	5.4	62
Eluate from immunoadsorbent[a]	16	3.2	970	300	860	44.5

[a]Immobilised anti-bovine kidney GGT (light form) antibody (two 5 ml columns).
Elution of GGT with 0.1 M ethanolamine in 0.5% Triton X-100.

regenerated immunoadsorbents can be used several times without significant loss in binding properties.

4.6. Examples

4.6.1 *Bovine Kidney GGT*

Light GGT form can be purified from bovine kidney by the general procedure described in Section 4.5. Increased specific activity can be achieved by re-chromatography of the enzyme eluted from an immunoadsorbent column (immobilised anti-bovine kidney GGT antibody). The product shows a 2500-to 3500-fold increase in GGT activity compared with the homogenate. The yield of purified enzyme amounts to 46%. All purification steps are summarised in *Table 2* and the elution profile from an immunoadsorbent column is presented in *Figure 9*.

Heavy form GGT is purified from bovine kidney using the following method.

(i)   Solubilise the GGT with 0.5% (v/v) Triton X-100 for 3 h at room temperature and centrifuge the suspension at 100 000 g for 60 min.
(ii)   Purify the supernatant on a single immunoadsorbent column (immobilised anti-bovine kidney GGT light form antibody).
(iii) Wash the column with PBS and then with 0.1 M ethanolamine in 0.5% (v/v) Triton X-100.

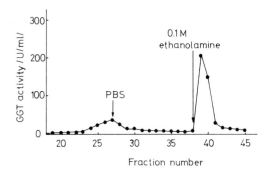

Figure 9. Elution profile of bovine kidney GGT (light form) from immunoadsorbent-immobilised anti-bovine GGT antibody. 140 ml of the supernatant (4000 U) after digestion of 100 000 g pellet with bromelain was passed through an immunoadsorbent column (1.2 x 4.5 cm) and 5 ml fractions collected (flow-rate 50 ml/h). The column was washed with PBS and finally with 0.1 M ethanolamine (10 ml/h).

Table 4. Purification of GGT (Light Form) from 400 g of Bovine Liver.

Preparation	Volume (ml)	Protein (mg)	Total activity (U)	Specific activity (U/mg)	Purification (-fold)	Yield (%)
Homogenate	1400	58 650	1953	0.033	1	100
100 000 g pellet suspension	360	19 800	1407	0.071	2.1	72
After solubilisation with bromelain	310	3100	1116	0.36	10.8	57
Eluate from immunoadsorbent[a]	32	0.89	625	750	22 730	32

[a]Immobilised anti-bovine kidney GGT antibody (5 ml column).
Elution of GGT with 0.1 M ethanolamine.

The purified preparation shows an 860-fold increase in GGT activity. Total yield of the purification amounts to 44.5%. The purification steps are summarised in *Table 3*.

4.6.2 *Bovine Liver GGT*

The light form of the enzyme can be purified from bovine liver by the general procedure described above. Increase of the enzyme activity is about 23 000 times and the total yield 32%. Purification data are presented in *Table 4*.

4.6.3 *Cow Milk GGT*

The light form of GGT can be purified on immunoadsorbent (immobilised anti-bovine kidney GGT antibody).

(i)   Adjust the pH of fresh skimmed milk to 4.6 with 1.0 M acetic acid and mix at room temperature for 1 h.

(ii)   Remove the precipitated casein by centrifugation (3000 g for 15 min) and centrifuge the resultant whey at 100 000 g for 60 min.

(iii) Collect the pellet and digest with bromelain.

Table 5. Purification of GGT (Light Form) from Cow Milk.

Preparation	Volume (ml)	Protein (mg)	Total activity (U)	Specific activity (U/mg)	Purification (-fold)	Yield (%)
Skim milk	1800	46 800	4030	0.086	1	100
Whey	1500	9900	3000	0.30	3.5	74
100 000 g pellet suspension	46	270	1980	7.35	85.5	49
After solubilisation with bromelain	43	194	1815	9.37	109	44
Eluate from immunoadsorbent[a]	19	0.57	617	1090	12 675	15

[a]Immobilised anti-bovine kidney GGT antibody (two 5 ml columns).
Elution of GGT with water.

Table 6. Purification of GGT (Light Form) from 72 g of Rat Kidney.

Preparation	Volume (ml)	Protein (mg)	Total activity (U)	Specific activity (U/mg)	Purification (-fold)	Yield (%)
Homogenate	250	2520	10 080	4	1	100
100 000 g pellet suspension	27	158	6650	42	10.5	66
After solubilisation with bromelain	25	66	6250	95	23.8	62
Eluate from immunoadsorbent[a]	24	3.5	4680	1320	330	46

[a]Immobilised anti-rat kidney GGT antibody (two 5 ml columns).
Elution of GGT with water.

The solubilised milk GGT may be further purified on immunoadsorbent as described in detail in Section 4.5. The enzyme eluted with water shows about 13 000 times higher activity than skimmed milk. Total yield for this purification is 15%. Purification data are presented in *Table 5*.

4.6.4 *Rat Kidney GGT*

Light GGT form can be purified from rat kidney by the general procedure described above using immobilised anti-rat kidney GGT antibody. The enzyme eluted with water shows 330 times higher activity than starting homogenate and the total yield is 46%. Purification steps are presented in *Table 6*.

4.6.5 *Mouse Kidney GGT*

Using a sensitive electrophoretic method for separation of a mixture of GGT-antibody a strong immunoreaction between mouse kidney GGT and anti-rat kidney GGT antibody can be demonstrated (32). This suggests that immobilised anti-rat kidney GGT antibody can be used as immunoadsorbent for purification of mouse kidney GGT. Purification data for mouse kidney enzyme are presented in *Table 7*. Purification details are as in the general procedure described above. After re-chromatography the increase of GGT activity is about 530 times and the total yield is 35%.

Table 7. Purification of GGT (Light Form) from 17 g of Mouse Kidney.

Preparation	Volume (ml)	Protein (mg)	Total activity (U)	Specific activity (U/mg)	Purification (-fold)	Yield (%)
Homogenate	65	2340	2500	12.1	1	100
100 000 g pellet suspension	40	900	2160	2.4	2.2	86
After solubilisation with bromelain	34	365	2000	5.8	5.1	80
Eluate from immunoadsorbent[a]	14	2.9	1360	475	432	54
After re-chromatography[a]	12	1.5	870	500	535	35

[a]Immobilised anti-rat kidney GGT antibody (two 5 ml columns).
Elution of GGT with 0.1 M ethanolamine.

Table 8. Purification of GGT (Light Form) from 180 g of Morris Hepatoma 5123D.

Preparation	Volume (ml)	Protein (mg)	Total activity (U)	Specific activity (U/mg)	Purification (-fold)	Yield (%)
Homogenate	800	17 200	640	0.04	1	100
100 000 g pellet suspension	150	3820	384	0.1	2.5	60
After solubilisation with bromelain	130	1315	290	0.2	5	46
Eluate from immunoadsorbent[a]	4	0.2	150	740	18 500	23

[a]Immobilised anti-rat kidney GGT antibody (1.5 ml column).
Elution of GGT with water.

4.6.6 *Rat Hepatoma GGT*

Using the general procedure described above, the light GGT form can also be purified from Morris hepatoma 5123D. The hepatoma cells are taken from Buffalo rats 24 days after implantation of a small sample of hepatoma pulp into hind leg (33). From immunoadsorbent (immobilised anti-rat kidney GGT antibody) the hepatoma enzyme may be eluted with water. Pure enzyme shows 18 500 times higher activity than the starting homogenate and total yield is 23%. Purification data are presented in *Table 8*.

4.7 **Final Remarks**

In double immunodiffusion tests, the GGT preparation purified from bovine kidney by immunoaffinity chromatography gives, with homologous antibody, three discrete precipitin arcs, all of which show GGT activity. The enzyme preparation may also be separated into four active fractions by isoelectric focusing. Rat kidney GGT purified on immunoadsorbent with homologous antibody gives two precipitin arcs possessing enzyme activity. These observations confirm the suggestion that GGT occurs in several molecular forms which may differ in sialic acid contents. Tate and Meister were the first to demonstrate that pure light GGT form from rat kidney can be separated into several enzymatic forms (34). These observations provide evidence that molecular forms purified on immunoadsorbents are immunologically indistinguishable.

5. AFFINITY CHROMATOGRAPHY ON HEPARIN-SEPHAROSE AND CASEIN-SEPHAROSE FOR THE PURIFICATION AND SEPARATION OF DNA-DEPENDENT RNA POLYMERASES AND PROTEIN KINASES

by G.Muszynska, E.Ber and G.Dobrowolska

5.1 Introduction

Although heparin is a common ligand for affinity chromatography, its full potential has still to be assessed. Heparin is an acidic mucopolysaccharide which interacts with basic surfaces of many proteins, and its linear structure may mimic the structure of nucleic acids. The ability of heparin to inhibit enzymes involved in nucleic acid metabolism has been used for purification on heparin-agarose of DNA and RNA polymerases (35 – 42), restriction endonucleases (43) and factors of protein synthesis (44). Recently it has been shown (45) that heparin-agarose may be utilised for simultaneous separation and purification of different types of protein kinases and various classes of DNA-dependent RNA polymerases. During this operation RNA polymerase I was eluted with casein kinase. For separation of these enzymes, casein covalently bound to Sepharose has been used. Casein kinases are those kinases which phosphorylate seryl and threonyl residues at acidic regions of the surfaces of proteins. Because casein is not a natural substrate of these enzymes, the biological role of these kinases remains unclear. There are suggestions that some of them might take part in phosphorylation of RNA polymerases (35), influencing the regulation of nucleic acid synthesis.

In this section methods are described using immobilised heparin and casein for the purification and separation of two classes of RNA polymerases and different types of protein kinases.

5.2 Basic Experimental Procedures

5.2.1 *Preparation of Enzyme Extracts*

Soak maize seedlings for 12 h; grow for 48 h in the dark at 26°C and harvest apical parts of seedlings. This material may be stored in liquid nitrogen.

For the enzyme extraction apply the following procedure.

(i)   Carefully powder frozen seedlings (20 g) with glass beads (3 – 4 g) in a pestle and mortar with a small amount of liquid nitrogen.

(ii)   After evaporation of liquid nitrogen extract the resulting powder by slowly adding 40 ml buffer containing 50 mM Tris-HCl, pH 7.9, 6 mM $MgCl_2$, 0.05% (w/v) monothioglycerol, 10 mM 2-mercaptoethanol, 0.1 mM phenylmethylsulphonyl fluoride (to inhibit protease activity) and 250 mM ammonium sulphate.

(iii)   Centrifuge the homogenate at 10 000 r.p.m. for 10 min.

(iv)   Keep the supernatant in an ice bath, resuspend the pellet in 20 ml of buffer and sonicate 3 x 40 sec.

(v)   Centrifuge the suspension as before and combine the supernatants from both spins.

Table 9. Buffers used for Affinity Chromatography Procedures.

Buffer	*Composition*
Buffers for heparin-Sepharose	
Buffer A	50 mM Tris-HCl, pH 7.9, 6 mM MgCl$_2$, 0.1 mM phenylmethylsulphonyl fluoride, 10 mM mercaptoethanol, 0.05% (v/v) monothioglycerol and 20% (v/v) glycerol
Buffer B	As Buffer A plus 75 mM ammonium sulphate
Buffer C	As Buffer A plus 300 mM ammonium sulphate
Buffer D	As Buffer A plus 600 mM ammonium sulphate
Buffers for casein-Sepharose	
Buffer E	10 mM Tris-HCl, pH 7.5, 10 mM MgCl$_2$, 22 mM NH$_4$Cl, 10 mM mercaptoethanol and 5% glycerol
Buffer F	As Buffer E plus 100 mM sodium chloride
Buffer G	As Buffer E plus 200 mM sodium chloride
Buffer H	As Buffer E plus 600 mM sodium chloride

(vi)   Precipitate supernatant proteins from this solution by slow addition of solid ammonium sulphate up to 2.2 M, stirring the solution continuously using a magnetic bar.

(vii)   After 1 h collect the precipitate by centrifugation at 5000 r.p.m. for 45 min.

(viii) Homogenise the ammonium sulphate pellet in a Potter homogeniser with buffer A (*Table 9*), sediment at 10 000 r.p.m. for 15 min and save the supernatant fraction; use this material as the starting point for enzyme purification.

5.2.2 Determination of Enzymatic Activities and Protein Content

Protein kinase assay. Make up in a final volume of 0.1 ml; 25 mM Tris-HCl, pH 7.9, 20 mM MgCl$_2$, 0.5 mM 2-mercaptoethanol, 1 mg/ml casein, 20 μM [^{32}P]-ATP (100–200 c.p.m./pmol). Start the reaction by addition of 40 μl of protein kinase preparation. Incubate for 20 min at 37°C; terminate the reaction by spotting 75 μl aliquots onto Whatman 3 MM disc (diameter 2.5 cm) and immediately wash five times with 5% (w/v) trichloroacetic acid (a total of 50 ml per disc), followed by 5 ml of 96% (v/v) ethanol. Dry the filters and count radioactivity using a toluene-based scintillator.

RNA polymerase assay. Make up in a final volume of 0.1 ml; 100 mM Tris-HCl, pH 8.0, 1 μM dithiothreitol, 10 mM MgCl$_2$, 1 mM MnCl$_2$, 25 μg heat-denatured calf thymus DNA, 50 nmol each of GTP, CTP and ATP and 5 μCi [^{3}H]UTP (130 pmol). Start the reaction by the addition of RNA polymerase preparation. After incubation for 20 min at 30°C, terminate the reaction by transferring 75 μl aliquots onto Whatman DE-81 disc paper. Wash the filters five times with 5% (w/v) cold trichloroacetic acid containing 1% (w/v) sodium pyrophosphate, and once with ethanol:ether (1:1), dry and count the radioactivity using a toluene-based scintillator.

Protein determinations. Bradford's Coomassie Blue method (46) was employed using bovine serum albumin as standard.

5.3 Equipment and Affinity Media for Affinity Chromatography

(i)   The only equipment needed consists of two chromatography columns (1.6 x 20 cm and 0.9 x 30 cm) which can be purchased from Pharmacia Fine Chemicals, (K16/20 and K9/30, respectively), a peristaltic pump (for pumping solutions onto the column) and a fraction collector.

(ii)   Cyanogen bromide-activated Sepharose 4B may be purchased from Pharmacia Fine Chemicals or Sepharose may be activated according to Axen *et al.* (47); see Section 1 of Chapter 1.

(iii)   Heparin-Sepharose may be obtained directly from Pharmacia Fine Chemicals, or may be prepared by coupling heparin with CNBr-activated Sepharose 4B as described by Iverius (48).

(iv)   Casein-Sepharose 4B may be obtained by coupling casein with CNBr-activated Sepharose 4B according to the following procedure (49); suspend casein (500 mg) in 0.1 M $NaHCO_3$, containing 0.5 M NaCl (250 ml). Because casein does not dissolve in this buffer, raise the pH to 13 with NaOH and then re-adjust to pH 8.0 with HCl. Mix this casein solution with CNBr-activated Sepharose 4B (50 g) and stir gently for 18 – 20 h at 4°C. Wash the suspension with 0.1 M $NaHCO_3$ containing 0.5 M NaCl and suspend in 1.0 M ethanolamine (pH 8.0), stirring, for 2 h. Wash twice with 0.1 M ethanolamine (pH 8.0), stirring, for 2 h. Wash twice with 0.1 M acetate buffer, pH 4.0, containing 1.0 M NaCl, and finally with 0.1 M borate buffer, pH 8.0, containing 1.0 M NaCl. Before packing, equilibrate the casein-Sepharose with Buffer E (*Table 9*).

5.4 Procedure for Heparin-Sepharose Chromatography

5.4.1 *Column Technique*

(i)   Into a column (1.6 x 20 cm) at 4°C pour 14 g of moist heparin-Sepharose suspended in Buffer B (~ 14 ml).

(ii)   Leave the gel to settle by 3 – 4 cm.

(iii)   Allow 12 – 13 ml of buffer to flow out of the column, then equilibrate the gel with Buffer B (150 ml) at a flow-rate of 16 – 20 ml/h, using a peristaltic pump.

(iv)   Quickly adjust the conductivity of the initial enzyme solution to that of Buffer B by dilution with Buffer A, to give a final volume of 100 – 110 ml.

(v)   Pump the enzyme solution onto the column and wash the column with 250 ml Buffer B (preferably overnight).

(vi)   Change to Buffer C and collect the fraction between 15 and 25 ml containing the peak of protein.

(vii)   Wash the column with Buffer C (~ 50 ml) until the protein content drops to less than 0.1 mg/ml. Elute with Buffer D; collect the protein peak which should emerge between 13 and 28 ml. This procedure allows separation of

    (a) histone kinase (eluted in the void)

    (b) casein kinase and a minor fraction of RNA polymerase, eluted by Buffer C, and

    (c) the main fraction of RNA polymerase – about 60% of the total activity, eluted by Buffer D (*Table 10* and *Figure 10*).

Table 10. Summmary of RNA Polymerases and Protein Kinase Purification.

Purification step	Histone kinase		Casein kinase		RNA polymerase		Protein mg
	Specific activity	Yield %	Specific activity	Yield %	Specific activity	Yield %	
Initial homogenate	50	— [a]	75	— [a]	1.5	— [a]	436
Ammonium sulphate precipitate	88	100	135	100	5	100	326
Heparin-agarose: flow through fraction	102	77	51	18	0.7	9	216
Eluted 300 mM ammonium sulphate	65	15	1662	87	15	21	23
Eluted 600 mM ammonium sulphate	110	2	393	3	251	61	4
Casein-agarose[b]: flow through fraction	ND	ND	110	4	21	17	16
Eluted 600 mM sodium chloride	ND	ND	38 100	57	0	0	0.7

[a]Due to the presence of nucleases and phosphatases, the total activity of initial homogenate was considerably lower.

[b]Fraction eluted 300 mM ammonium sulphate from heparin-agarose was applied on a casein-agarose column.

One unit of protein kinase activity was defined as the amount of enzyme that catalyses the transfer of 1 pmol of terminal group [^{32}P]ATP to the protein substrate/min at 37°C. One unit of RNA polymerase activity corresponds to incorportion of 1 pmol of UMP into RNA/min at 30°C. Specific activities of the investigated enzymes were defined as units/mg protein.

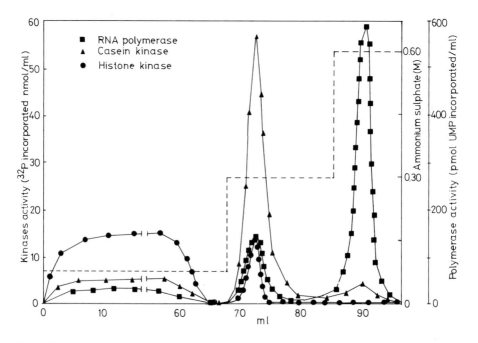

Figure 10. Separation on heparin-Sepharose column of RNA polymerase and protein kinases.

If a linear salt gradient is used instead of a step gradient, two peaks of RNA polymerase activity are still obtained, indicating that this is not due either to overloading of the column or an inappropriate choice of step gradient.

5.4.2 *The Combination of Batch and Column Procedures for Heparin-Sepharose*

To speed up the most time-consuming steps of the procedure (binding of the enzymes to heparin-Sepharose and washing the column) a batch procedure may be applied for the first stages of the separation, and might also be applicable to scaling up the enzyme purification. The procedure involves the following operations.

(i)   Equilibrate heparin-Sepharose (20 g) with Buffer B to appropriate conductivity.

(ii)   Quickly adjust the conductivity of the initial enzyme solution to that of Buffer B, by dilution with Buffer A.

(iii)   Pour the enzyme solution onto the equilibrated heparin-Sepharose and leave for 1 h, stirring gently, and centrifuge the suspension for 5 min at 3000 r.p.m.

(iv)   Collect the supernatant, pour Buffer B (70 ml) onto the gel and shake gently to mix. Centrifuge the suspension for 5 min at 3000 r.p.m. and discard the supernatant.

(v)   Wash the gel with 3 x 70 ml of Buffer B.

(vi)   Suspend the gel in about 15 ml of Buffer B, pour into a column (1.6 x 20 cm) and leave to settle by about 5 cm.

(vii)   Allow 15 ml of buffer solution to flow out from the column, and pump Buffer C onto the column using a peristaltic pump at a flow-rate of 16 – 20 ml/h.

(viii) Collect the fraction 13 and 23 ml of eluate containing the peak of protein.

This batch procedure reduces the laboratory work from 2 days to one, and allows the same separation of enzymes as in the case of the column technique. However, the recovery and the degree of purification is about 30% lower than using the column technique.

5.5 **Procedure for Casein-Sepharose Chromatography**

The fraction eluted from the heparin-Sepharose column by Buffer C contains the majority of casein kinase activity and some of the RNA polymerase activity (*Figure 10*). Chromatography on casein-Sepharose may be used to separate and further purify these enzymes. The removal of excess salt from the above fraction (containing 300 mM ammonium sulphate) is essential; this may be achieved by overnight dialysis in Buffer E (3 x 500 ml).

(i)   After dialysis, centrifuge the sample for 5 min at 10 000 r.p.m. and discard the precipitate.

(ii)   Fill a column (0.9 x 30 cm) with wet agarose (10 g) at 4°C and allow the gel to settle by about 3 cm.

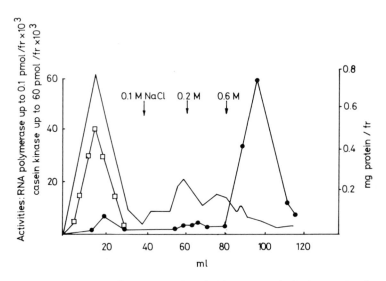

Figure 11. Separation of RNA polymerase I and casein kinase on casein-Sepharose 4B column (———) protein concentration; (□) RNA polymerase and (●) casein kinase.

(iii) Allow 8 ml of buffer solution to flow out of the column, then pump 100 ml of Buffer E through at a flow-rate of 16 – 20 ml/h.

(iv) Apply the dialysed enzyme solution to the column and collect the first volume through the column.

(v) Wash with Buffer F (20 ml), Buffer G (20 ml), then change to Buffer H and collect the eluate between 16 and 35 ml of Buffer H.

Under these conditions RNA polymerase is not bound to the affinity column and is eluted in the void, whereas strongly bound casein kinase is released from the column by the high ionic strength Buffer H (*Figure 11*). Chromatography achieves the separation of kinase and polymerase activities, with a considerable purification of high specific activity casein kinase (*Table 10*).

Note: the packing and column equilibration should be carried out the day before the separation step.

5.6 Characterisation of RNA Polymerase Fractions Released from Heparin-Sepharose Column

The first fraction of RNA polymerase is eluted from the column with Buffer C containing 300 mM ammonium sulphate and the second with Buffer D containing 600 mM ammonium sulphate. The main criteria for RNA polymerase classification include sensitivity to (i) α-amanitin and (ii) order of elution from ion exchangers (50). The first fraction of RNA polymerase is resistant to α-amanitin at concentrations up to 400 μg/ml, wherease the second fraction is inhibited by 60% at a concentration of 0.1 μg/ml. By comparison with published data (50,51), the first fraction seems to correspond to RNA polymerase I and the second fraction to RNA polymerase II. This is confirmed by the order of elution (using a linear salt gradient) from DEAE Sephadex. Also the ratio of the activities eluted at 300 mM and 600 mM ammonium sulphate indicates that these fractions represent nucleolar RNA polymerase I and nucleoplasmic polymerase II, respectively.

5.7 Concentration and Storage of Enzymatic Preparations

The proportions of affinity support to applied protein (14 g heparin-Sepharose to protein from 20 g of maize seedlings) are optimal for this separation. Using smaller quantities of protein, the enzyme eluted is too dilute, whereas using more protein extract results in prolonged column steps and partial inactivation of enzyme. If the recommended conditions are used subsequent concentration of enzyme should not be necessary. At this stage the enzyme is stable and can be stored for several weeks (at $-70°C$) without loss of activity, or it can be used directly for the next step of purification.

Casein kinase may be concentrated, if necessary, in an Amicon filtration chamber equipped with a UM 10 membrane, and should be stored in small portions in liquid nitrogen at a protein concentration of $0.5 - 1.0$ mg/ml.

5.8 Regeneration and Storage of Affinity Supports

The above affinity supports can be used several times without detectable loss of binding capacity. After use, strongly bound impurities may be removed by washing the column as follows:

(i)   2.0 M ammonium sulphate (2 volumes);
(ii)   1.0 M Tris, pH 8.0, containing 6.0 M urea (3 volumes);
(iii) distilled water (5 volumes).

Store at 4°C in equilibration buffer, containing sodium azide (0.02% w/v).

5.9 Notes on the Experimental Procedures

All steps of enzyme extraction, purification and concentration should be performed at $2-4°C$. Due to instability of the reducing agents monothioglycerol and mercaptoethanol, these should be incorporated into the buffers immediately prior to use. CNBr-activated Sepharose 4B is unstable and must be used immediately for coupling with ligands. α-Amanitin is an extremely toxic compound and special precautions must be taken when working with it. The exact ionic strength of the preparation applied to the affinity columns is critical for the separation and purification of the enzymes.

6. COUPLING OF *m*-AMINOPHENYLBORONIC ACID TO TRIAZINE-ACTIVATED SUPPORTS

Longstaff (52) has investigated the use of four different buffers for optimal coupling of *m*-aminophenylboronic acid (aPBA) to triazine-activated agarose beads. Suspend moist activated gel (2 g) in buffer (2 ml) and to this add buffer (4 ml) containing 0.1 M aPBA. Mix the suspension at room temperature on a Coulter mixer, and remove aliquots at intervals, wash and assay for boron. Results are shown in *Figure 12*. Many workers recommend using buffers of high

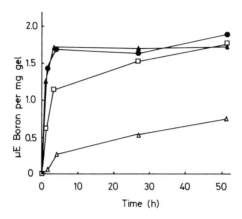

Figure 12. Time course of the coupling of *m*-aminophenylboronic acid to triazine-activated Matrex Gel. Coupling solutions; 10 mM NaOH (●), 0.5 M K phosphate, pH 7.5 (▲), 0.5 M Tris-HCl, pH 8.8 (□), and 0.5 M Na bicarbonate-carbonate, pH 10.25 (△).

pH when coupling amino-containing ligands to various types of activated supports, so that the amine group will be available in its uncharged form for nucleophilic attack on the activated matrix. However, the pK_a of aromatic primary amino groups is much lower than that of aliphatic amine groups, hence milder conditions may be used to couple aPBA to triazine-activated gels. The results in *Figure 12* show that coupling proceeds rapidly at high pH (10 mM NaOH) but is equally rapid at the lowest pH tested (potassium phosphate, pH 7.5). The slow reaction observed with bicarbonate buffer, pH 10.25, is interesting because many workers recommend the use of bicarbonate buffers for immobilisation of various ligands to many types of activated gels. It might be widely assumed that if a specific set of coupling conditions is useful for one particular application, they will also work well in other instances. However, these results suggest otherwise.

Coupling in Tris-HCl, pH 8.3, proceeds at an intermediate rate but once again the gel eventually contains as much boron as those gels coupled in either NaOH or phosphate buffer. Therefore, Tris does not itself react with the activated gel. The lack of reactivity of Tris may be further demonstrated by the lack of reaction after incubating activated gel with 2.0 M Tris at pH 10.0 and at 60°C for 18 h. The gel is able to immobilise equal amounts of aliphatic diamine before and after the treatment and the total nitrogen content of the gel is unchanged following incubation. A related compound, 2-amino-2-methyl-1,3-propanediol, is also found to be unreactive towards triazine-activated gels. From structural models of these two molecules, it can be seen that, in the most stable form, the nitrogen is sterically hindered by hydroxymethyl groups which may severely restrict its nucleophilicity. To demonstrate this it can be shown that neither of these two amines reacts with ninhydrin to produce a blue product. For these reasons it seems likely that Tris would be equally unreactive towards other activated matrices of various types, and thus could be used in coupling buffers without any problem. In fact, the use of Tris in coupling solutions is widely discouraged, and it is often recommended as being useful to block remaining underivatised groups following ligand immobilisation.

For coupling aPBA to other supports e.g., triazine-activated paper, epoxy-activated Sepharose 6B (Pharmacia) and Eupergit C (Rohm Pharma), 0.5 M potassium phosphate buffer, pH 7.8, containing 0.1 M aPBA has been found to give satisfactory results. Excess active groups may be blocked using ethanolamine (1.0 M).

6.1 **Chromatography using Immobilised PBA**

(i) *Nucleosides*. Use either 1.5 ml (3.9 x 0.7 cm i.d.) or 2 ml (5.2 x 0.7 cm i.d.) columns with a flow-rate of 2 ml/h, collecting either 1 ml or 2 ml fractions. Equilibrate the column with running buffer (50 mM potassium phosphate, containing 1.0 M NaCl, pH 7.8) for about 24 h before use. Apply a mixture of nucleosides and deoxynucleosides (2.5 μmol of each) to the column in running buffer (total volume 0.4 ml). Allow the applied solution to drain into the gel, just below the surface, before washing any remaining sample into the gel with a further aliquot of buffer (0.4 ml). Develop the column with running buffer. Collect deoxynucleotides in the void volume and desorb tightly-bound nucleosides with 100 mM sorbitol in buffer. Columns may be run at 4°C or room temperature. Record elution profiles by measuring absorbances at 260 nm.

(ii) *Glycoproteins*. Apply protein (2 – 4 mg) in approximately 250 μl of running buffer (50 mM taurine/NaOH pH 8.7, containing 20 mM MgCl$_2$ – this concentration may need to be adjusted to between 0 and 50 mM) to a column (2 ml) of PBA-Matrex Gel. Develop the column at a flow-rate of 2 ml/h, collecting 2 ml fractions. Elution of bound protein is effected using running buffer containing 50 mM sorbitol or 50 mM Tris/HCl.

Columns may be regenerated by washing with 50 mM NaOH or 200 mM sodium acetate buffer pH 5.4.

7. APPLICATIONS OF IMMOBILISED DYES

7.1 **Preparation of Dye Matrices**

7.1.1 *Method of Heyns and De Moor* (53)

(i) Pour agarose beads (5 g moist weight) onto a sintered glass funnel and wash with distilled water under suction to remove the azide preservative.

(ii) Continue suction gently until the surface of the gel cracks.

(iii) To this agarose add a solution of the appropriate dye (80 mg in 2.5 ml of distilled water) and mix on a Coulter mixer.

(iv) After addition of Na$_2$CO$_3$ solution (168 mg in 2.5 ml of distilled water) and thorough mixing, incubate the suspension for 40 h at 45°C with occasional stirring.

(v) Pour the dyed gel into a sintered glass funnel and wash with distilled water until no more free dye is removed.

(vi) Wash the gel with 1.0 M KCl and 4.0 M urea to remove any further dye before extensive washing with distilled water.

7.1.2 *Method of Dean and Watson* (54)

Some dye matrices may be prepared more efficiently by this 'salting in' procedure. The addition of salt to the incubation mixture leads to faster reaction times with less hydrolysis of the triazine moiety (58).

(i)   To agarose beads (10 g moist weight) in distilled water (25 ml) add a solution of the dye (250 mg in 25 ml of distilled water) and rotate on a Coulter mixer for 5 min.

(ii)   Add a 20% (w/v) solution of NaCl and mix for a further 30 min.

(iii)   At the end of this period add 1.0 M Na_2CO_3 (12.5 ml) and stir the mixture thoroughly before incubating for 40 h at 45°C with occasional stirring.

(iv)   After coupling, wash the dyed gel free of unbound dye as described above.

Dye matrices of different ligand concentrations may be prepared by varying the amount of dye added to the gel in the coupling procedure.

7.1.3 *Storage and Regeneration of Dye Matrices*

Dye matrices should be stored wet at 4°C in the presence of 0.1% (w/v) sodium azide when not in use. After use, dye matrices may be regenerated by washing with 4.0 M urea in 0.5 M NaOH (three column volumes) to remove bound material before re-equilibrating the column with the appropriate buffer (15 – 20 column volumes).

7.1.4 *Dye Screening Procedure*

Dye screening is conducted at 4°C.

(i)   Pack dye matrices into polycarbonate columns (2.3 cm x 0.7 cm).

(ii)   Equilibrate the columns with 10 – 15 column volumes of equilibration buffer before applying the sample which has been previously dialysed against equilibration buffer.

(iii)   Wash the columns with equilibration buffer until a total of 5 ml has been collected.

(a) If biospecific elution is to be attempted, dissolve the appropriate eluant in equilibration buffer (1 ml), re-adjust the pH and apply the eluant to the column. Wash with equilibration buffer until a total of 5 ml has been collected.

(b) In the case of a non-specific elution method, wash the column with equilibration buffer containing 1.0 M KCl.

Bunemann and Muller (55) have developed a cross-linked bisacrylamide gel to which nucleotide base-specific dyes are covalently attached by spacers of polyacrylamide chains of different lengths.

Immobilised (AT)-specific malachite green or (GC)-specific phenyl neutral red are used for base pair-specific fractionation of DNA from bacterial or mammalian sources, and can be used to separate defined fragments of phage DNA.

Structure-specific gels with immobilised acridine yellow ED or phenyl neutral red can also be used for specific separations of supercoiled DNA from other native, open circular, or linear forms of DNA (56).

7.2 General Procedure for Chromatography of DNA on Dye-substituted Bis-acrylamides

7.2.1 *Pre-treatment of the Gel and Packing of the Chromatographic Column*

(i) Suspend the gel (20 g) in approximately 10 volumes of 10 mM sodium phosphate buffer, containing 1 mM EDTA, pH 6.0.

(ii) After the gel has settled, decant the slightly turbid supernatant and resuspend the material in the same buffer.

(iii) After repeating this procedure at least three times, transfer the material into a chromatography column of suitable size (1.2 x 22 cm column, ~ 25 ml) and wash with 2.0 M sodium perchlorate in 10 mM sodium phosphate buffer, pH 6.0, containing 1 mM EDTA.

(iv) Equilibrate the column with two column volumes of 10 mM sodium phosphate buffer, pH 6.0, containing 1 mM EDTA (2 – 3 column volumes). The 2.0 M perchlorate wash removes particles of dye-substituted gel (which impair column flow).

7.2.2 *Affinity Chromatography*

(i) The DNA to be loaded onto the column must be pre-purified to be free of contaminating protein.

(ii) Dissolve the DNA in 10 mM sodium phosphate buffer, containing 1 mM EDTA, pH 6.0 to give a concentration of $50 - 100$ μg ($1 - 2$ A_{260} units) DNA/ml and load onto the column. The maximum capacity of the gels amounts to approximately 40 μg (0.8 A_{260} units) DNA/ml settled gel.

(iii) After the column has been washed several times with the same buffer, elute the DNA with a linear gradient of sodium perchlorate, $0 - 2.0$ M in 10 mM sodium phosphate buffer, pH 6.0, containing 1 mM EDTA. Ideally, use flow-rates of $10 - 15$ ml/h and collect fractions of $5 - 8$ ml.

In some cases, when eluting supercoiled DNA from structure-specific gels, it may be necessary to raise the perchlorate concentration to 5.0 M in sodium phosphate buffer, pH 6.0, containing EDTA.

7.2.3 *Regeneration and Storage*

The affinity adsorbent columns can be regenerated and re-used indefinitely. Regenerate the gels by washing with at least 2 column volumes of 5.0 M sodium perchlorate, 10 mM sodium phosphate buffer, pH 6.0, containing 1 mM EDTA, followed by several column volumes of 10 mM sodium phosphate buffer containing 1 mM EDTA, pH 6.0. Store the columns in the dark at room temperature.

Important note: Care must be taken that the regeneration of the malachite green gel is carried out at pH 6.0 or lower. At pH values above 6.0, the malachite green base will convert to the colourless carbinol base, which has neither affinity nor specificity for DNA (57).

The affinity adsorbent gels, especially the phenyl neutral red gel, are extremely sensitive to light. Care must therefore be taken not to expose the columns to light, even during chromatography, e.g., by wrapping the columns in aluminium foil.

7.2.4 *Applications*

The (AT)-specific malachite green gel and the (GC)-specific phenyl neutral red gel can be used for base pair-specific affinity chromatography of double-stranded DNA. For base-specific separation of DNA choose the gel from which the desired DNA fraction is eluted first, i.e., for the isolation of (AT)-rich satellite DNA, the phenyl neutral red gel (GC-specific) is used, and for the isolation of (GC)-rich calf thymus DNA use the malachite green gel (AT-specific).

Successful applications of malachite green gel include the separation of satellite and chromosomal DNA from *Halobacterium halobium* (59) and the separation of the triplexes poly (dA)·(dT)·(rU) and poly (dA)·2(rU) (63), whereas the red gel has been used in the preparation of size-fractionated nematode DNA (60) for cloning experiments.

The resolving power of this method for the affinity chromatography of nucleic acids is in most cases superior to DNA fractionation in buoyant density gradients in the presence of base pair-specific ligands.

The acridine yellow ED gel and the phenyl neutral red gel are suitable for use for the structure-specific separation of supercoiled DNA from open circular or linear DNA, because these dyes form intercalating complexes with the energetically favoured supercoiled DNA (56,62).

A rapid method for the preparation of DNA from *Escherichia coli* containing the plasmid pBR322 by acridine yellow affinity chromatography has been described (61).

8. IMMOBILISATION OF NUCLEOTIDES

8.1 Preparation of N^6-(6-Aminohexyl)-5'-AMP (64)

(i) In a glass ampoule (155 x 15 mm) place a solution of the sodium salt of 6-mercaptopurine riboside 5'-monophosphate (I, *Figure 13*) (10 mg, 0.024 mmol) and 1,6-hexanediamine (*Figure 13*, II, n = 6) (200 mg, 1.75 mmol) in water (0.6 ml).

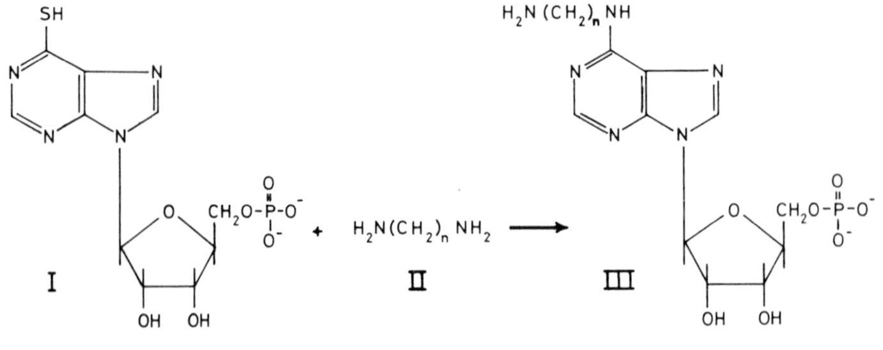

Figure 13. Reaction in the preparation of N^6-(6-aminohexyl)-5'-AMP.

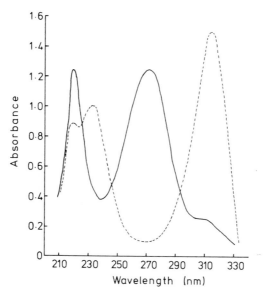

Figure 14. Absorption spectra at pH 12 of 6-mercaptopurine riboside 5′-phosphate (10 mg) and hexanediamine (200 mg) in water (0.6 ml). (– – –) initial spectrum, (———) spectrum after 16 h at 85 – 95°C (64).

(ii)   Seal the ampoule and heat overnight (16 h) in a horizontal position (N.B., the reaction appears to depend on the ratio of surface to volume of ligand) at 85 – 95°C in an aluminium heating block.

(iii)   The course of the reaction may be followed by spectral analysis of a suitable dilution in water. The reaction is characterised by a decreased absorption at 309 nm and a concomitant increase at 267 nm (*Figure 14*). This decrease is characteristic of the change in the maximum absorbance between 6-mercaptopurine and N^6-alkyl-substituted aminopurine at alkaline pH. Complete reaction is indicated by the absence of absorption at 309 nm. Quantitative conversion is usually effected in 16 h, although in some instances a further period of heating is required.

(iv)   After cooling, dilute the sample with water (10 ml) and lyophilise to yield a pale-yellow hygroscopic residue.

(v)   Repeat this procedure several times to remove excess hexanediamine.

(vi)   Dissolve the final straw-coloured residue in a minimum volume of water and apply at pH 10.0 to a column (5 x 50 mm) of Dowex 1-X2 (formate) resin, 200 – 400 mesh at room temperature.

(vii)   Wash the column with 0.1 M ammonium formate, pH 7.0 to remove impurities followed by elution of the nucleotide with a linear gradient of ammonium formate pH 5.5 (0.1 – 1.0 M, 50 ml total volume).

(viii)   Collect 1 ml fractions and pool the fractions with positive absorbance at 267 nm.

(ix)   Lyophilise and remove the residual ammonium formate by vacuum sublimation, to a constant weight, over a 3:1 (w/w) mixture of NaOH and anhydrous $CaCl_2$.

8.2 **Coupling of N⁶-(6-Aminohexyl)-5′-AMP to Sepharose 4B**

(i)   Activate 10 g (moist weight) of Sepharose 4B with cyanogen bromide.

(ii)   Wash the gel with ice-cold 0.1 M NaHCO₃ buffer pH 10.0, and add to a solution of N⁶-(6-aminohexyl)-5′-AMP (10 mg) in 0.1 M NaHCO₃, pH 10.0 (5 ml).

(iii) Rotate the suspension overnight on a Coulter mixer at 4°C.

(iv)   Wash the matrix thoroughly with water, 1 M KCl and water; store in sodium azide (0.02% w/v).

9. HYDROPHOBIC INTERACTION CHROMATOGRAPHY

Hydrophobic interaction chromatography provides a general systematic approach to the purification of both water-soluble and lipophilic proteins. The technique makes use of the presence of hydrophobic sites exposed on the surface of proteins, which results in differing degrees of hydrophobic interactions with an uncharged matrix containing hydrophobic groups.

The approach described below makes use of homologous series of alkyl agaroses and their derivatives (*Figure 15*) to achieve resolution and purification of proteins and cells (65).

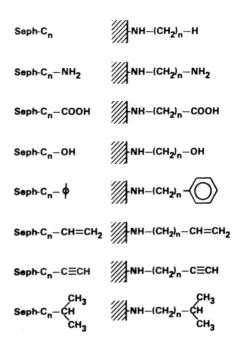

Figure 15. A variety of homologous series of alkylagarose derivatives (Seph-C$_n$-X) which can be used for hydrophobic interaction chromatography. Additional series can be formed by using other functional groups, by placing the functional group at different positions along the hydrocarbon chain, or by having more than one functional group on the chain (65).

9.1 Synthesis of Sepharose-C$_n$ (n = 1 − 12) by the CNBr Procedure

(i)   Activate Sepharose 4B by reaction with CNBr at pH 10.5 − 11.0 and 22°C.

(ii)   Add CNBr (1 g dissolved in 1 ml of dioxane) to agarose (10 g wet weight) suspended in water (20 ml).

(iii)   Initiate activation by raising the pH of the mixture to 11.0 with 5.0 M NaOH.

(iv)   Allow the reaction to proceed for 5 min with gentle swirling, maintaining the pH between 10.5 and 11.0 with 5.0 M NaOH.

(v)   Terminate activation by addition of crushed ice, filtration and washing of the gel with ice-cold water (100 ml).

(vi)   Dissolve the appropriate n-alkylamine (4 mol/mol of CNBr used for the activation of the agarose) in a solvent composed of equal volumes of N,N′ dimethylformamide and 0.1 M NaHCO$_3$ (pH 9.5) (20 ml) and adjust the pH, if necessary, to 9.5. With the higher members in the series (n > 8) a gel or a precipitate may form; the mixture should be warmed to about 60°C to obtain a clear solution, cooled rapidly and mixed with the activated agarose before precipitation recurs.

(vii) Allow the reaction to proceed at room temperature (20 − 25°C) for 10 h, with gentle swirling.

(viii) Filter the alkylagarose and wash with five volumes of each of the following: distilled water, 0.2 M acetic acid, distilled water, 20 mM NaOH, water, dioxane-water (1:1 v/v), 50 mM acetic acid, and finally with about 20 volumes of distilled water.

(ix)   Before use, wash and equilibrate each column with the buffer chosen for chromatography.

Alternative procedure (66)

(i)   Carry out the coupling step in dioxane:water (95:5 v/v); allow the reaction to proceed for 20 h at pH 9.0, 22°C.

(ii)   Wash the column materials successively with 2 − 3 volumes of each of the following dioxane-water mixtures (v/v) 95:5; 80:20; 60:40; 40:60; and 20:80.

(iii)   Transfer the resulting gel into 0.1 M NaHCO$_3$, pH 9.5, shake for 1 h at room temperature, then wash with about 20 volumes of water and suspend in the chosen buffer.

The same synthetic procedure should be used throughout the series in order to make the columns identical in all respects except for the length of their alkyl side chains. In systematic studies aimed at establishing a standard purification procedure, it is essential that the ligand and charge density of the column materials be determined as outlined below.

9.2 Synthesis of Aminoalkyl Sepharose-C$_n$-NH$_2$ (n = 2 − 12) by the CNBr Procedure

Sepharose-C$_n$-NH$_2$ represents Sepharose 4B activated by CNBr and reacted with an α,ω-diaminoethane *n* carbon atoms long. This homologous series is prepared by the procedure described above for the alkyl Sepharose-C$_n$ series.

9.3 **Properties and Characterisation of the Column Materials**

9.3.1 *Determination of Dry Weight*

Wash the gel with 20 volumes of deionised water, weigh three 1 g samples (wet weight), and dry them under reduced pressure at 120°C to constant weight. Take the average of the triplicate as the dry weight of that particular column material.

9.3.2 *Determination of the Ligand Density of Columns with Radioactively Labelled Ligands*

(i)   Dissolve samples of column materials (0.3 g, wet weight) in 1 ml of 6.0 M HCl, and add 10 ml of Triton-toluene scintillator.

(ii)   Shake the mixture until a clear solution is formed then count in a scintillation counter [scintillation liquid composed of 1,4-bis-[2-(5-phenyl-oxazolyl)] benzene (0.1 g), 2,5-diphenyloxazole (8 g), toluene (666 ml) and Triton X-100 (333 ml)].

(iii) Determine the efficiency of radioactivity counting by counting six identical samples of the radioactively-labelled free ligand (e.g., n-[1-^{14}C]dodecyl alcohol), three samples alone and three in the presence of 0.3 g (wet weight) of the appropriate non-radioactively-labelled column material dissolved each in 1 ml of 6.0 M HCl as described above.

(iv)   Use the experimentally determined counting efficiency to correct the values obtained in the ligand density determination.

9.4 **Selecting a Column for a Given Purification**

The selection of a suitable substituted agarose within a given homologous series, as in the alkyl Sepharose series, is achieved by the use of a group of column materials designed for this purpose. This 'kit' is composed of a set of Pasteur pipettes (0.5 x 5 cm), each containing a different member of the homologous series, e.g., Sepharose-C_n, where n = 1 – 10. In addition, two control columns are included: one of unmodified agarose (Seph-C_O) and the other (Seph-C_{ON}) of agarose which has been activated with CNBr and then treated with ammonia in an attempt to simulate, in part, the functional groups introduced by the activation with CNBr and the subsequent coupling with an amine. A similar kit of column materials can be prepared for any other series of the Seph-C_n-X type.

When aliquots of a protein mixture (e.g., an extract of muscle) are applied to each of this group of columns, some of the proteins become adsorbed and some are not bound. If the mixture of unadsorbed protein from each of the columns is then subjected to polyacrylamide gradient gel electrophoresis in the presence of SDS, it can be shown (65) that increasing the hydrocarbon chain length results in the binding of more proteins, indicating that a greater number of proteins are adsorbed onto the columns.

9.5 **Optimising the Separation of Two Enzymes**

For the resolution of two enzymes, it is necessary to obtain the characteristic adsorption profile of each, by monitoring their concentration in the unbound fractions. If the two differ in their adsorption profiles, it is usually possible to select a column which retains one enzyme while not binding the other. For example, when an extract of rabbit muscle is subjected to chromatography on Seph-C_n-NH$_2$, the columns with short chains (n = 2 or 3) do not bind glycogen synthetase I, but retain the enzyme on Seph-C_4-NH$_2$, from which it can be eluted with a linear NaCl gradient. A column of Seph-C_n-NH$_2$ retains synthetase I but does not bind glycogen phosphorylase b, which in this column series and under the same conditions requires hydrocarbon chains five and six carbon atoms long for retention. Therefore it is possible to isolate glycogen synthetase by passing muscle extract through Seph-C_n-NH$_2$, and then to extract phosphorylase b by subjecting the unbound proteins to chromatography on Seph-C_6-NH$_2$, setting the stage for the principle of consecutive fractionation described below.

In a preparative experiment (67), 70 ml of rabbit muscle extract containing 1.75 g protein and a synthetase activity of 0.2 units/mg of protein were applied to a Seph-C_4-NH$_2$ column (2.4 x 10 cm) equilibrated with a buffer composed of 50 mM sodium β-glycerophosphate, 50 mM 2-mercaptoethanol and 1 mM EDTA, pH 7.0. After the unadsorbed protein was removed by washing, a linear NaCl gradient in the same buffer was applied. Glycogen synthetase I was thus purified 20- to 50-fold in one step. Since Seph-C_4-NH$_2$ discriminates very efficiently between this enzyme and haemoglobin, glycogen synthetase could be isolated from human erythrocytes with a purification factor of 600-fold in one step.

9.6 Selecting the Series

The exploratory kits of columns are very useful for deciding not only which column in the series is likely to provide optimal purification, but also in deciding upon the most suitable series of column materials. The resolution of clostripain and collagenase (68) is a case in point. Both enzymes have very similar adsorption profiles on the Seph-C_n series (n = 1 − 7), both being retained on Seph-C_6. However, they exhibit different adsorption profiles with the Seph-C_n-NH$_2$ series; whereas clostripain is essentially retained already on Seph-C_4-NH$_2$, collagenase is not bound by this column. Under the conditons of the experiment, a higher member in the series (Seph-C_7-NH$_2$) is required to completely retain the enzyme. Moreover, the binding of the enzymes to Seph-C_4-NH$_2$ and to Seph-C_7-NH$_2$ is quantitatively reversed in both cases when 1.0 M NaCl is included in the eluant.

On the basis of this exploratory experiment it is clear that it is advisable to resolve the two enzymes by passage of the crude preparation through Seph-C_4-NH$_2$ in order to extract clostripain, and then to extract collagenase by applying the unadsorbed proteins to Seph-C_7-NH$_2$. This consecutive use of two columns from the Seph-C_n-NH$_2$ series brings about resolution and purification of both enzymes.

9.7 Elution Methods

Systematic studies aimed at optimising protein elution from alkylagaroses have indicated a variety of influencing factors, including polarity-reducing agents, specific 'deformers', mild detergents, concentration of denaturants, alterations in pH, temperature, ionic strength and buffer composition. Since the availability of hydrophobic sites on the surface of a protein appears to depend on its conformation, and since the retention of proteins by alkylagaroses depends largely on the lipophilicity, size, shape and distribution of these sites, the means of elution described above may function either by directly disrupting the hydrophobic interactions between the column material and the protein, by changing the conformation of the protein, or by a combination of both.

High selectivity in elution may also be achieved by the use of biospecific ligands such as coenzymes, substrates, specific metal ions and allosteric effectors which often bring about ligand-induced conformational changes in proteins.

Glycogen phosphorylase *b* can be eluted from Seph-C$_4$ by polarity-reducing compounds such as ethylene glycol (69). Upon increasing the concentration of this glycol in the elution solvent to 30% (v/v), the enzyme can be desorbed from the same column by simply lowering the pH of the eluting buffer to 5.6, which is known to bring about distinct structural changes in this enzyme. Addition of urea (1.0 M) to the irrigating buffer also promotes desorption of the enzyme. However, from the point of view of preserving the native conformation of this enzyme (highest catalytic activity and unaltered response to regulatory metabolites) the optimal eluant was found to be 0.4 M imidazole citrate, pH 7.0.

The term 'deformer' was introduced to describe a compound that brings about a localised, rather limited conformational change in a protein, a change which can be fully and readily reversed by removal of the compound. Different proteins will have different specific 'deformers', and numerous examples of enzymes which are exceptionally sensitive to specific cations, anions or uncharged compounds are recorded in the literature. Their effect on the structure of the enzyme is often reflected in its reversible inactivation (affecting the K_m or V_{max} or both), its aggregation state, or its binding constant for certain ligands. Very often, such compounds are mentioned in reporting the optimisation of the assay procedure for a given enzyme, and referred to as non-competitive inhibitors, whose presence in the assay medium is to be avoided. Some of these compounds may act as specific 'deformers' and should therefore be screened with the exploratory kit to determine their effectiveness as eluants.

9.8 Applications of Hydrophobic Interaction Chromatography

9.8.1 *Use of Phenyl-Sepharose in the Purification of Lymphocyte Colony Enhancing Factor (LCEF) (70)*

(i)   Use a crude extract obtained by ammonium sulphate precipitation of conditioned media derived from a culture of mitogen-stimulated peripheral blood mononuclear cells.

(ii) Collect the supernatant after saturating to 40% with ammonium sulphate, followed by centrifugation, dialyse against 20 mM phosphate buffer, pH 6.8, and bring to 4.0 M NaCl by the addition of solid NaCl.

(iii) Load onto a column (1.3 x 3 cm), previously equilibrated with Buffer A (4.0 M NaCl in 20 mM phosphate, pH 6.8).

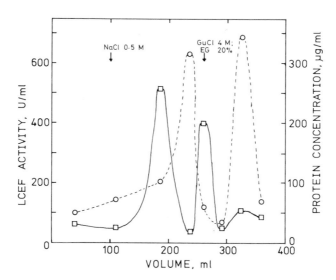

Figure 16. Partial purification of LCEF from human T-lymphocytes (74). (-----) LCEF activity, (———) protein concentration.

(iv)   Wash the column with Buffer A (80 ml), and elute with Buffer B (0.5 M NaCl in 10 mM phosphate, pH 7.1) (80 ml) at a flow-rate of 16 – 18 ml/h.

(v)   Apply 80 ml of the viscous Buffer C (4.0 M guanidine hydrochloride in 20% ethylene glycol, pH 12.0); this reduces the flow-rate to 7 – 8 ml/h.

(vi)   Finally, wash the column with 50% (v/v) aqueous ethanol and water before equilibrating with buffer.

(vii)   Pool active fractions (as determined by measurement of protein concentration) and concentrate either by lyophilisation or by ultrafiltration on a UM-10 Diaflo membrane (in the case of the 4.0 M NaCl fractions). The guanidinium-containing fractions need to be dialysed against phosphate buffer prior to concentration.

(viii) Dialyse concentrated fractions exhaustively against phosphate buffer before assaying for LCEF activity (see *Figure 16*). Measure LCEF activity by a two-step method of colony formation of peripheral blood leucocyte responder cells (70).

9.8.2 *Hydrophobic Chromatography of Enzymes from Halophilic Bacteria using Decreasing Concentration Gradients of Ammonium Sulphate*

(i)   Harvest cells from approximately 6 litres of anaerobic culture of *Halobacterium* after 4 days and pack by centrifugation.

(ii)   Suspend the cell pellet in 4.3 M NaCl in 10 mM phosphate, pH 7.3, and sonicate for 4 min at 0°C in a Branson sonicator equipped with a microtip.

(iii)   Centrifuge the sonicated suspension for 1 h in an ultracentrifuge at 80 000 *g* (30 000 r.p.m.) at 25°C.

(iv)   Dialyse the supernatant against 1.6 M ammonium sulphate in 50 mM sodium phosphate, pH 6.6, and centrifuge as before.

(v)   Add solid ammonium sulphate to the supernatant to a final concentration of 2.5 M.

(vi)   Stir for 1 h, centrifuge the suspension above and collect the supernatant.

Column chromatography. Chromatography is carried out using Sepharose 4B, CM-cellulose, HMD-agarose, DEAE-cellulose or Celite (71). All operations may be carried out at room temperature

(i)   Pack a column with gel and wash with several volumes of 2.5 M ammonium sulphate in 50 mM phosphate buffer, pH 6.6.

(ii)   Apply the bacterial extract onto the column and wash with about one column volume of 2.5 M ammonium sulphate.

(iii)   Apply a linear gradient decreasing from 2.5 M to 0.5 M ammonium sulphate and elute at a flow-rate of 18 – 35 ml/h. Assay alternate fractions for glutamate dehydrogenase and malate dehydrogenase. Typical results are shown in *Table 11* and *Figure 17*.

10. USE OF IMMOBILISED INHIBITORS IN THE CHROMATOGRAPHY OF PEPSINS

10.1 Preparation of Affinity Sorbents

10.1.1 *Low Concentration Immobilised Inhibitor*

(i)   Dissolve ϵ-aminocaproyl-L-phenylalanyl-D-phenylalanyl methyl ester (250 mg) in the minimum volume of dimethylformamide, and add triethylamine (76 μl) and Separon H1000 (hydroxyalkylmethacrylate gel) modified with epichlorohydrin (4 g, epoxide group content 800 μmol/g).

(ii)   Shake the mixture for 48 h, filter, wash with water until neutral, and then with ethanol and diethyl ether.

(iii)   Wash the product with 6.0 M guanidine hydrochloride solution and water, dry for analysis to constant weight at 105°C and transfer for affinity chromatography into the respective buffer. (At the original tripeptide concentrations in solution of 0.02, 0.04, 0.12 and 0.25 mol/l the dried product contains 0.85, 1.2, 2.5, and 4.5 μmol/g of affinity ligand, respectively).

Table 11. Elution of two Halophilic Enzymes by Decreasing Concentration Gradients of Ammonium Sulphate[a]

Column	GDH(M)	MDH(M)
Sepharose 4B	1.44	1.70
CM-cellulose	1.84	2.06
HMD-agarose	1.18	
DEAE-cellulose	ne[b]	ne[b]
Celite	3.11	3.53
Solubility[c]	3.01	3.32

[a]Ammonium sulphate concentration at which the enzymes eluted from the various solid supports.
[b]Not eluted: the enzymes did not elute until 0.4 M $(NH_4)_2SO_4$ (even at 0.3 M $(NH_4)_2SO_4$ in the case of GDH). They could, however, be eluted by a NaCl gradient.
[c]Ammonium sulphate concentration at which 50% of the activity was found in the supernatant.

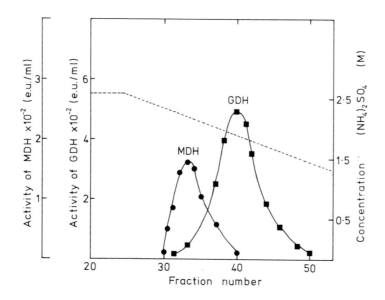

Figure 17. Ammonium sulphate fractionation of glutamate dehydrogenase (■) and malate dehydrogenase (●) from *Halobacterium* using CM-cellulose (71).

10.1.2 *High Concentration Immobilised Inhibitor*

(i)   Dissolve ε-aminocaproyl-L-phenylalanyl-D-phenylalanine methyl ester (11.1 g) in a small volume of methanol and make up to 75 ml with Britton-Robinson buffer (pH 11.0).

(ii)   Shake 30 ml of this solution with 5 g of dry gel (Separon H300, epoxide group content 600 μmol/g), filter the suspension and wash the gel with water, 6.0 M guanidine hydrochloride solution and finally water.

(iii)   Further treatment is as in Section 10.1.1. The content of bound inhibitor is 155 μmol/g of dry carrier.

10.2 Chromatography of Porcine Pepsin on ε-Aminocaproyl-L-phenylalanyl-D-phenylalanyl-methyl-Separon Columns with Various Concentrations of Immobilised Inhibitor

Apply a solution of pepsin (1 g/200 ml) in 0.1 M acetate buffer, pH 4.5, to a column (9 x 0.8 cm) of ε-aminocaproyl-L-Phe-D-Phe-OCH$_3$-Separon prepared as described in Section 10.1, and equilibrated with 0.1 M sodium acetate solution, pH 4.5. Wash the column with the equilibration buffer and subsequently desorb pepsin with 0.1 M acetate buffer, pH 4.5, containing 1.0 M sodium chloride. Typical results for porcine pepsin are shown in *Figure 18*.

11. ISOLATION OF VITAMIN B$_{12}$-BINDING PROTEINS USING AFFINITY CHROMATOGRAPHY

Allen and Majerus (73,74) have isolated vitamin B$_{12}$-binding proteins using affinity chromatography on vitamin B$_{12}$-Sepharose. Vitamin B$_{12}$ derivatives may be

Ligands for Immobilisation

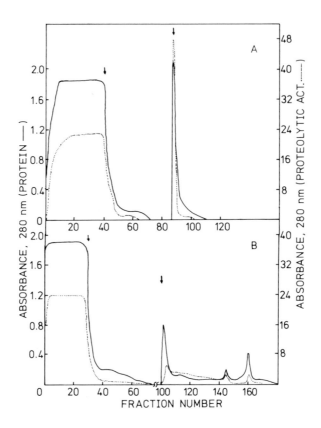

Figure 18. Affinity chromatography of porcine pepsin on ε-aminocaproyl-L-Phe-D-Phe-OCH₃-Separon columns with **(A)** low and **(B)** high concentrations of immobilised inhibitor (72). Inhibitor concentration; **(A)** 0.85 μmol/g dry support, **(B)** 155 μmol/g dry support. (————) protein concentration; (-----) proteolytic activity.

prepared by partial acid hydroysis (in 0.4 M HCl for 64 h at room temperature) of the amide groups of the unsubstituted propionamide side chains of the corrin ring of vitamin B_{12}. The resultant mixture of mono-, di- and tri-carboxylic vitamin B_{12} derivatives may be separated by chromatography on QAE-Sephadex.

11.1 Coupling of Vitamin B_{12} Derivatives to 3,3′-Diaminodipropylamine-substituted Sepharose

(i)   Dissolve the lyophilised derivatives of vitamin B_{12} in water to give a vitamin B_{12} derivative concentration of 10.7 μmol/ml.

(ii) Add 30 ml of this solution with 4.3 ml of water to a beaker containing packed 3,3′-diaminodipropylamine-substituted Sepharose (30 ml) and mix with a magnetic stirrer at 22°C.

(iii) Increase the pH from 3.9 to 5.6 with 0.2 M NaOH (~ 1.2 ml); add 3.0 ml of l-ethyl-3-(3-diethylaminopropyl)carbodiimide, 50 mg/ml, in aliquots of 0.6 ml at 1 min intervals, stirring continuously.

(iv)　　Continue gentle stirring at room temperature (22°C) in the dark for 18 h.

(v)　　Collect the substituted Sepharose by vacuum filtration in a Buchner funnel and wash at 22°C successively with distilled water (300 ml), 0.1 M glycine-NaOH, pH 10.0 (600 ml), distilled water (300 ml) and 0.1 M potassium phosphate, pH 7.0 (300 ml).

The washed substituted Sepharose is deep red in colour and contains 0.68 μmol of vitamin B_{12} per ml of packed Sepharose.

12. REFERENCES

1.　Porath,J. (1974) in *Methods in Enzymology*, Vol. **34**, Jakoby,W.B. and Wilchek,M. (eds.), Academic Press Inc. London and New York, p.13.
2.　Gribnau,T.C.J., van With,T., van Sommeren,A., Roeles,F., van Hell,H. and Schuurs,A. (1979) *INSERM (Colloquium on Affinity Chromatography)*, **86**, 175.
3.　Porath,J., Carlsson,J., Olsson,I. and Belfrage,G. (1975) *Nature*, **258**, 298.
4.　Sundaram,P.V. and Hornby,W.E. (1970) *FEBS Lett.*, **10**, 325.
5.　Cuatrecasas,P. (1970) *J. Biol. Chem.*, **245**, 3059.
6.　Weetall,H.H. (1969) *Nature*, **223**, 959.
7.　Alebic-Kolbah,T., Rendic,S., Fuks.Z., Sunjic,V. and Kajfez,F. (1979) *Acta Pharm. Jugoslav.*, **29**, 53.
8.　Stewart,K.K. and Doherty,R.F. (1973) *Proc. Natl. Acad. Sci. USA*, **70**, 2850.
9.　Lagercrantz,C., Larsson,T. and Karlsson,H. (1979) *Anal. Biochem.*, **99**, 352.
10.　Lagercrantz,C., Larsson,T. and Denfors,I. (1981) *Comp. Biochem. Physiol.*, **69C**, 375.
11.　Allenmark,S. (1982) *Chem. Scr.*, **20**, 5.
12.　Allenmark,S., Bomgren,B. and Boren,H. (1982) *J. Chromatogr.*, **237**, 473.
13.　Allenmark,S. and Bomgren,B. (1982) *J. Chromatogr.*, **252**, 297.
14.　Allenmark,S., Bomgren,B. and Boren,H. (1983) *J. Chromatogr.*, **264**, 63.
15.　Bristow,P.A., Brittain,P.N., Riley,C.M. and Williamson,B.F. (1977) *J. Chromatogr.*, **131**, 57.
16.　Allenmark,S. (1984) *J. Biochem. Biophys. Methods*, **9**, 1.
17.　Allenmark,S., Bomgren,B. and Andersson,S. (1984) *Prep. Biochem.*, **14**, 139.
18.　Allenmark,S., Bomgren,B., Boren,H. and Lagerstrom,P.-O. (1984) *Anal. Biochem.*, **136**, 293.
19.　Curthoys,N.P. and Hughey,R.P. (1979) *Enzyme*, **24**, 383.
20.　Tate,S.S. and Meister,A. (1981) *Mol. Cell. Biochem.*, **39**, 357.
21.　Stromme,J.H. and Theodorsen,L. (1976) *Clin. Chem.*, **22**, 417.
22.　Lowry,O.H., Rosenbrough,N.J., Farr,A.L. and Randall,R.J. (1951) *J. Biol. Chem.*, **193**, 265.
23.　Dulley,J.R. and Grieve,P.A. (1975) *Anal. Biochem.*, **64**, 136.
24.　Davis,B.J. (1964) *Ann. N.Y. Acad. Sci.*, **121**, 404.
25.　Prusak,E., Siewinski,M. and Szewczuk,A. (1980) *Clin. Chim. Acta*, **107**, 21.
26.　Ouchterlony,O. (1949) *Acta Pathol. Microbiol.*, **26**, 507.
27.　Szewczuk,A. and Baranowski,T. (1963) *Biochem. Zeit.*, **338**, 317.
28.　Milnerowicz,H., Prusak,E., Siewinski,M., Ziomek,E., Kustrzeba-Wojcicka,I. and
29.　Szewczuk,A. (1981) *Arch. Immunol. Ther. Exp.*, **29**, 543.
30.　Tate,S.S. and Meister,A. (1975) *J. Biol. Chem.*, **250**, 4619.
31.　Livingston,D.M. (1974) in *Methods in Enzymology*, Vol. **34B**, Jakoby,W.B. and Wilchek,M. (eds.), Academic Press Inc., New York, p.723.
32.　March,S.C., Parikh,J. and Cuatrecasas,P. (1974) *Anal. Biochem.*, **60**, 149.
33.　Prusak,E., Kurowska,E. and Szewczuk,A. (1983) *Arch. Immunol. Ther. Exp.*, **31**, 613.
34.　Albert,Z., Szewczuk,A. and Albert,W. (1977) *Neoplasma*, **24**, 49.
　　Tate,S.S. and Meister,A. (1976) *Proc. Natl. Acad. Sci. USA*, **73**, 2599.
35.　Sternbach,H., Engelhardt,R. and Lezius,A.C. (1975) *Eur. J. Biochem.*, **60**, 51.
36.　Teissere,M., Penon,P., Azou,Y. and Ricard,J. (1977) *FEBS Lett.*, **82**, 77.
37.　Jaehning,J.A., Woods,P.S. and Roeder,R.G. (1977) *J. Biol. Chem.*, **252**, 8762.
38.　Smith,S.S. and Braun,R. (1978) *Eur. J. Biochem.*, **82**, 309.
39.　Zillig,W., Stetter,K.O. and Tobien,M. (1978) *Eur. J. Biochem.*, **91**, 193.
40.　Spindler,S.R., Duester,G.L., D'Alessio,J.M. and Paule,M.R. (1978) *J. Biol. Chem.*, **253**, 4669.
41.　Spindler,S.R., D'Alessio,J.M., Duester,G.L. and Paule, M.R. (1978) *J. Biol. Chem.*, **253**, 6242.
42.　Brennessel,B.A., Buhrer,D.P. and Gottlieb,A.A. (1978) *Anal. Biochem.*, **87**, 411.
43.　Bickle,T.A., Pirotta,V. and Imber,R. (1977) *Nucleic Acids Res.*, **4**, 2561.

44. Waldman,A.A., Marx,G. and Goldstein,J. (1975) *Proc. Natl. Acad. Sci. USA,* **72**, 2352.
45. Ber,E., Muszynska,G., Tarantowicz-Marek,E. and Dobrowolska,G. (1983) in *Proceedings of the Fifth International Symposium on Affinity Chromatography and Biological Recognition, Annapolis,* Chaiken,I.M., Wilchek,M. and Parikh,J. (eds.), Academic Press, New York, in press.
46. Bradford,M.M. (1976) *Anal. Biochem.,* **72**, 248.
47. Axen,H. and Ernback,S. (1971) *Eur. J. Biochem.,* **18**, 351.
48. Iverius,P.H. (1971) *Biochem. J.,* **124**, 677.
49. Rahmsdorf,H.J. and Pai,Sh.-H. (1979) *Biochim. Biophys. Acta,* **267**, 339.
50. Chambon,P. (1975) *Annu. Rev. Biochem.,* **44**, 613.
51. Jendrisak,J. and Guilfoyle,T.J. (1978) *Biochemistry (Wash.),* **17**, 1322.
52. Longstaff,C. (1983) Ph.D. Thesis, University of Liverpool.
53. Heyns,W. and deMoor,P. (1974) *Biochim. Biophys. Acta.,* **358**, 1.
54. Dean,P.D.G. and Watson,D.H. (1979) *J. Chromatogr.,* **165**, 301.
55. Bünemann,H. and Müller,W. (1978) *Nucleic Acids Res.,* **5**, 1059.
56. Bünemann,H. and Müller,W. (1978) in *Affinity Chromatography,* Hoffmann-Ostenhof,O. *et al.,* (eds.), Pergamon Press, New York, p.353.
57. Koller,B., Delius,H., Bünemann,H. and Müller,W. (1978) *Gene,* **4**, 227.
58. Beech,W.F. (1970) *Fibre-reactive Dyes,* published by Logos Press, London.
59. Wiedinger,G., Klotz,G. and Goebel,W. (1979) *Plasmid,* **2**, 377.
60. Karn,J., Brenner,S., Barnett,L. and Cesarchi,H. (1980) *Proc. Natl. Acad. Sci. USA.,* **77**, 5172.
61. Vincent,W.S. and Goldstein,E.S. (1981) *Anal. Biochem.,* **110**, 123.
62. Shatle,D., Koshland,D., Weinstock,G.H. and Bornstein,D. (1980) *Proc. Natl. Acad. Sci. USA,* **77**, 5375.
63. Bryngelsson,T. and Pero,W.R. (1980) *Acta Chem. Scand.,* **60**, 456.
64. Craven,D.B., Harvey,M.J., Lowe,C.R. and Dean,P.D.G. (1974), *Eur. J. Biochem.,* **41**, 329.
65. Shaltiel,S. (1984) in *Methods in Enzymology,* Vol. **104**, 69.
66. Halperin,G., Breitenbach,M., Tauber-Finkelstein,M. and Shaltiel,S. (1981) *J. Chromatogr.,* **215**, 211.
67. Shaltiel,S. and Ev-El,Z. (1973), *Proc. Natl. Acad. Sci. USA,* **70**, 778.
68. Kula,M.R., Halef-Haghir,D., Tauber-Finkelstein,M. and Shaltiel,S. (1976) *Biochem. Biophys. Res. Commun.,* **69**, 389.
69. Shaltiel,S. (1975) *FEBS Proceedings,* **46**, 117.
70. Werber, M.M., Daphna,D., Goldman,J., Joseph,D., Rednay,J. and Rozenszajn,L.A. (1953) *Immunology,* **50**, 261.
71. Mevarech,M., Weicht,W. and Werber,M.M. (1976) *Biochemistry (Wash.),* **15**, 2383.
72. Turkova,J., Blaha,K. and Adanova,K. (1982) *J. Chromatogr.,* **236**, 375.
73. Allen,R.M. and Majerus,P.W. (1972) *J. Biol. Chem.,* **247**, 7695.
74. Allen,R.M. and Majerus,P.W. (1972) *J. Biol. Chem.,* **247**, 7702.

CHAPTER 6

Quantitative Characterisation of Interactions by Affinity Chromatography

DONALD J. WINZOR

1. INTRODUCTION

Despite the introduction of quantitative affinity chromatography a decade ago $(1-3)$, the relative infancy of the technique as a method for studying interactions is evident from the fact that there have been virtually as many investigations concerned with methodological developments as there have been reports of its application to experimental systems. Most of those studies have entailed the use of affinity chromatography to evaluate equilibria involving enzymes and modifiers, inhibitors or substrates $(1-12)$, but the method has also been used to comment upon protein-drug interactions (13), antigen-antibody systems (14) and, to a lesser extent, protein-protein interactions (15,16). Basically, the technique entails the immobilisation of a biospecific reactant group X on an inert chromatographic matrix (e.g., frequently Sepharose), and measurement of the weight-average elution volume ($\bar{V}_A$) of the partitioning solute A in a series of chromatography experiments in which the solute migrates in the presence of a range of concentrations of a ligand, S, that also interacts specifically with A and/or X. In some instances the resultant variation in $\bar{V}_A$ has reflected competition between soluble and immobilised reactants for the same A site $(2-5,7-14)$: in others it has reflected the interaction of immobilised reactant X with the binary AS complex formed between solute and soluble ligand (1,6). The aim of quantitative affinity chromatography is the interpretation of this variation of $\bar{V}_A$ in terms of the various operative equilibria. Since the positions of equilibria are concentration-dependent in accordance with the law of mass action, it is clearly essential (3,4) that frontal chromatography be employed for precise definition of the reactant mixture to which $\bar{V}_A$ refers.

2. BASIC THEORETICAL EXPRESSIONS

Initially, quantitative affinity chromatography was confined to the determination of association constants for systems in which the partitioning solute A possessed only a single site for interaction with X and/or S, but subsequent developments $(17-19)$ have extended the theory to take into account multivalency of the solute. Existing theory is, however, restricted either to systems in which no gel partitioning of solute occurs, or to systems where the ligand is sufficiently small to permit the approximation to be made that A and all AS_i complexes have identical gel-partitioning characteristics. It may be possible to ensure the former condition in situations where both solute and ligand are macromolecular by the suitable choice of an affinity matrix that excludes the larger reactant. Although the theory for multivalent solutes must obviously also include the special case of a univalent solute, the concepts of quantitative affinity chromatography are more clearly introduced by first considering the simplest situations.

2.1 Systems with a Univalent Partitioning Solute

A comprehensive investigation of the frontal chromatographic behaviour of affinity systems in which the partitioning solute A possessed at most a single site for each of X and S (3) considered the following equilibria to represent the possible interactions likely to be encountered in affinity chromatography.

$$A + S \rightleftharpoons AS; \qquad m_{AS} = K_{AS} m_A m_S$$

$$A + X \rightleftharpoons AX; \qquad m_{AX} = K_{AX} m_A m_X$$

$$X + S \rightleftharpoons XS; \qquad m_{XS} = K_{XS} m_X m_S$$

$$XS + A \rightleftharpoons XSA; \quad m_{XSA} = K_{XSA} m_A m_S m_X$$

$$AS + X \rightleftharpoons ASX; \quad m_{ASX} = K_{ASX} m_A m_S m_X$$

In these expressions the molar concentration of each species is assumed to be the same in all regions of the column that are accessible to the species, X being considered to be uniformly distributed throughout the same volume as A. From a general relationship encompassing affinity chromatography of a system in which all equilibria are operative, a series of expressions has been derived (3) for experimentally realistic situations in which three of the five equilibrium constants are zero. These relationships are summarised in *Table 1* in which V_A denotes the elution volume of A in an experiment conducted with A alone at the same concentration ($\bar{m}_A$) to which $\bar{V}_A$ refers; and V_A^* the elution volume in an experiment with no solute-matrix interaction ($V_A = V_A^*$ for cases 1 and 2). Should A interact with X (cases 3 and 4), this elution volume for unretarded solute may be obtained either from extrapolation of results to infinite concentration of competing ligand, S, or from an experiment with the same matrix material devoid of X groups (2,3,20). Alternative expressions for these two cases in terms of $(\bar{V}_A - V_A^*)$ have been included in *Table 1* to provide expressions for use with systems where V_A^* is more readily determined than V_A because of the impractically large value of the latter. In earlier versions (3,21) of *Table 1*, the volume of stationary phase, V_s, was substituted for V_A^* in the right hand side of these equations due to incorrect allowance for the volume in which matrix sites are considered to be distributed (22).

Several points are noteworthy in relation to *Table 1*.

(i) Cases 1 and 2 describe affinity chromatography systems in which the magnitude of $\bar{V}_A$ increases with increasing concentration of ligand, whereas cases 3 and 4 refer to competitive situations that lead to a decrease in elution volume with increasing m_S.

Table 1. Affinity Chromatography Systems for Which the Elution Volume of Univalent Solute, A, is Affected by Inclusion of a Ligand, S. Case operative equilibrium dependence of elution volume upon ligand.

Case	Operative equilibrium constants	Dependence of elution volume upon ligand concentration
1	K_{AS}, K_{ASX}	$\dfrac{1}{\bar{V}_A - V_A^*} = \dfrac{1}{V_A^* \bar{\bar{m}}_X K_{AS} K_{ASX} m_S} + \dfrac{1 + K_{ASX} \bar{m}_A}{V_A^* \bar{\bar{m}}_X K_{ASX}}$
2	K_{XS}, K_{XSA}	$\dfrac{1}{\bar{V}_A - V_A^*} = \dfrac{1}{V_A^* \bar{\bar{m}}_X K_{XS} K_{XSA} m_S} + \dfrac{1 + K_{XSA} \bar{m}_A}{V_A^* \bar{\bar{m}}_X K_{XSA}}$
3	K_{AS}, K_{AX}	$\dfrac{1}{\bar{V}_A - V_A} = \dfrac{1}{V_A^* \bar{\bar{m}}_X K_{AX} K_{AS} m_S} - \left[\dfrac{(1 + K_{AX} \bar{m}_A)^2}{V_A^* \bar{\bar{m}}_X K_{AX} K_{AS} m_S}\right] + \dfrac{1 + K_{AX} \bar{m}_A}{V_A^* \bar{\bar{m}}_X K_{AX}} \quad ; \quad \dfrac{1}{\bar{V}_A - V_A^*} = \dfrac{K_{AS} m_S}{V_A^* \bar{\bar{m}}_X K_{AX}} + \dfrac{1 + K_{AX} \bar{m}_A}{V_A^* \bar{\bar{m}}_X K_{AX}}$
4	K_{AX}, K_{XS}	$\dfrac{1}{\bar{V}_A - V_A} = \dfrac{1}{V_A^* \bar{\bar{m}}_X K_{AX} K_{XS} m_S} - \left[\dfrac{(1 + K_{AX} \bar{m}_A)^2}{V_A^* \bar{\bar{m}}_X K_{AX} K_{XS} m_S}\right] + \dfrac{1 + K_{AX} \bar{m}_A}{V_A^* \bar{\bar{m}}_X K_{AX}} \quad ; \quad \dfrac{1}{\bar{V}_A - V_A^*} = \dfrac{K_{XS} m_S}{V_A^* \bar{\bar{m}}_X K_{AX}} + \dfrac{1 + K_{AX} \bar{m}_A}{V_A^* \bar{\bar{m}}_X K_{AX}}$

(ii) Unique identification of the relevant case is possible provided that the concentration dependence of the elution volumes of the separate reactants (A and S) is also investigated.

(iii) Taken in conjunction with the concentration dependence of $\bar{V}_A$ at a fixed ligand concentration (m_S), the plots of $1/(\bar{V}_A - V_A)$ *versus* $1/m_S$ [or $1/(\bar{V}_A-V_A^*)$ *versus* m_S (cases 3 and 4)] suggested by the relationships of *Table 1* may be used to provide values of the operative equilibrium constants.

(iv) All relationships contain the total solute concentration in the liquid phase $\bar{m}_A$, a quantity that poses no problem in frontal chromatography, but one that is of unknown and ever-changing magnitude in zonal studies.

(v) The theoretical relationships also contain the parameter $\bar{\bar{m}}_x$ the total concentration of accessible matrix sites, the value of which may be substantially smaller than that inferred from chemical analysis of the affinity matrix.

2.2 Allowance for Multivalency of the Solute

In the event that the partitioning solute possesses more than one site for interaction with matrix sites the above chemical equilibria continue to be operative, but clearly it is necessary to consider additional equilibria for the formation of species AX_i in which one molecule of A forms cross-links between several matrix sites. Initial treatments of solute multivalency (3,12) were restricted to the limiting situation in which steric requirements precluded such multiple binding of A to the matrix. Under those special conditions, realisable experimentally with some systems by employing very low values of $\bar{\bar{m}}_x$ (6,14), plots in accord with the relationships in *Table 1* are linear and yield k_{AX}, the intrinsic association constant (23) or site-binding constant (24) for the relevant interaction between matrix and partitioning solute species (A or AS_i). However, the use of an affinity column with a higher concentration of matrix sites leads to curvilinearity of the suggested plots (14,17), which indicates non-fulfilment of the implicit assumption that multiple binding of solute to the matrix may be neglected.

By resorting to reacted-site probability theory (25 – 27), a general solution to this problem of multivalency of the partitioning solute has been obtained (18,19) for situations in which the several interactions of f-valent A with X continue to be described in terms of a single site-binding constant, k_{AX}. The original expressions (18,19) developed for partition equilibrium experiments were written in terms of the solute concentration in the liquid phase ($\bar{m}_A$) and its total concentration in the system ($\bar{\bar{m}}_A$), but may be converted to a form more appropriate to frontal chromatography on noting (22) that the relationship $V_A^* \bar{\bar{m}}_A = \bar{V}_A \bar{m}_A$ expresses the necessary condition of mass conservation. From *Table 2*, which lists the chromatographic set of expressions for the situations considered in relation to univalent solutes, it is evident that $\bar{\bar{m}}_x$, the effective total concentration of matrix sites, is again a parameter for which the magnitude must be deduced in order to achieve quantitative assessment of experimental results in terms of the two operative equilibrium constants.

Table 2. Affinity Chromatography Systems for Which the Elution Volume of f-Valent Solute, A, is Affected by Inclusion of a Ligand, S.

Case	Operative equilibrium constants		Dependence of elution volume upon ligand concentration
1	k_{AS}, k_{ASX}	$k_{ASX} =$	$\dfrac{(1 + k_{AS}m_S) [1 - (V_A^*/\bar{V}_A)^{1/f}]}{(V_A^*/\bar{V}_A)^{1/f}\{\bar{\bar{m}}_X - f\bar{m}_A(\bar{V}_A/V_A^*)[1 - (V_A^*/\bar{V}_A)^{1/f}]\}k_{AS}m_S}$
2	k_{XS}, k_{XSA}	$k_{XSA} =$	$\dfrac{(1 + k_{XS}m_S) [1 - (V_A^*/\bar{V}_A)^{1/f}]}{(V_A^*/\bar{V}_A)^{1/f}\{\bar{\bar{m}}_X - f\bar{m}_A(\bar{V}_A/V_A^*)[1 - (V_A^*/\bar{V}_A)^{1/f}]\}k_{XS}m_S}$
3	k_{AS}, k_{AX}	$k_{AX} =$	$\dfrac{(1 + k_{AS}m_S) [1 - (V_A^*/\bar{V}_A)^{1/f}]}{(V_A^*/\bar{V}_A)^{1/f}\{\bar{\bar{m}}_X - f\bar{m}_A(\bar{V}_A/V_A^*)[1 - (V_A^*/\bar{V}_A)^{1/f}]\}}$
4	k_{AX}, k_{XS}	$k_{AX} =$	$\dfrac{(1 + k_{XS}m_S) [1 - (V_A^*/\bar{V}_A)^{1/f}]}{(V_A^*/\bar{V}_A)^{1/f}\{\bar{\bar{m}}_X - f\bar{m}_A(\bar{V}_A/V_A^*)[1 - (V_A^*/\bar{V}_A)^{1/f}]\}}$

3. EXPERIMENTAL ASPECTS OF QUANTITATIVE AFFINITY CHROMATOGRAPHY

From the viewpoint that $\bar{\bar{m}}_X$, the total concentration of accessible matrix sites, is a parameter whose magnitude requires evaluation from the experimental results, column chromatography is the simplest procedure to employ for quantitative affinity chromatography, since the same value of $\bar{\bar{m}}_X$ then pertains to all results obtained with the same solute on the same column. However, in some instances, particularly those involving biological affinity matrices such as muscle myofibrils (28), the column technique is not feasible because of flow-rate problems; and accordingly partition equilibrium experiments also find application in quantitative affinity chromatography.

3.1 Column Chromatography

With the possible exception of a requirement for conducting experiments on a relatively small scale, the procedures for setting up quantitative affinity chromatography do not differ substantially from those for any other column experiment. However, the following points should be taken into consideration.

(i)   In view of the general dependence of chemical equilibria upon temperature, a jacketed column, thermostatically maintained at the temperature of interest, should be used.

(ii) Since evaluation of equilibrium constants requires the accurate assessment of elution volumes, precautions are necessary to ensure sufficient precision in the measured values of $\bar{V}_A$. If the column effluent is being divided into fractions for assay, the size of each fraction is most accurately determined from its weight: collection of fractions in previously weighed tubes is therefore recommended. Alternatively, an accurately controlled and measured flow-rate may be used in conjunction with continuous monitoring of the column effluent.

153

(iii)　As noted above in relation to *Table 1*, the quantitative expressions contain $\bar{m}_A$, the total concentration of partitioning solute in the mobile phase. Frontal chromatography should therefore be used in preference to the usual zonal procedure.

Frontal chromatography differs from its zonal counterpart only in regard to the volume of solution applied to the column. A sufficient volume of solution is applied to ensure that the elution profile contains a plateau region in which the concentrations of all soluble reactants equal those applied. As in the zonal method, elution with buffer is then recommenced. The elution pattern thus obtained is a combination of two separate profiles: an advancing pattern in which the concentrations of all species increase from zero to those in the applied mixture; and a trailing elution profile in which the concentrations of all species decrease from the plateau values to zero. In quantitative affinity chromatography, which usually employs an assay procedure specific for the total concentration of the partitioning solute, the constituent elution volume, $\bar{V}_A$, may be obtained as the median bisector, or centroid (29), of either the advancing or trailing pattern. Specifically, the median bisector is defined by the expression

$$\bar{V}_A = \frac{1}{\bar{m}_A^\alpha} \int_0^{\bar{m}_A^\alpha} V d\bar{m}_A \simeq \frac{1}{\bar{m}_A^\alpha} \Sigma V \Delta \bar{m}_A \qquad \text{Equation 1}$$

where V is the mean effluent volume corresponding to an increment in solute concentration $\Delta \bar{m}_A$, and the limits of the summation (trapezoidal integration) are the solvent plateau ($\bar{m}_A = 0$) and the plateau of original composition ($\bar{m}_A = \bar{m}_A^\alpha$). More convenient expressions for the evaluation of $\bar{V}_A$ from experiments in which the column effluent is divided into fractions are as follows:

$$\bar{V}_A = \frac{1}{\bar{m}_A^\alpha} \left[V' - \sum_0^{V'} \bar{m}_A \Delta V \right] \qquad ; \text{ advancing profile} \qquad \text{Equation 2a}$$

$$\bar{V}_A = \frac{1}{\bar{m}_A^\alpha} \left[\sum_0^{V''} \bar{m}_A \Delta V \right] \qquad ; \text{ trailing profile} \qquad \text{Equation 2b}$$

In these summations $\bar{m}_A$ is the total solute concentration in a fraction, the volume of which is ΔV, V' corresponds to an effluent volume located within the plateau region of original composition (where $\bar{m}_A = \bar{m}_A^\alpha$), and V'' to an effluent volume in the solvent plateau ($\bar{m}_A = 0$) that is regenerated in the trailing elution profile. In the advancing profile the origin of the volume scale coincides with commencement of sample application, whereas the trailing profile commences with the re-application of solvent to the column. It should be noted that a single frontal experiment yields two independent estimates of $\bar{V}_A$, the effluent volume at which the solute concentration is half of its plateau value if the boundary is symmetrical.

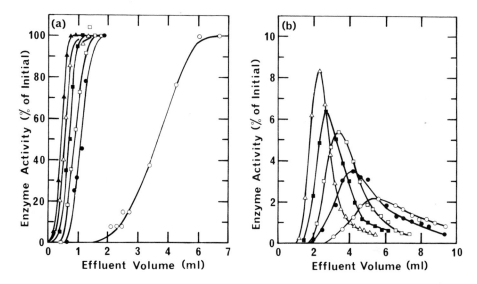

Figure 1. Comparison of the effects of NADH concentration on the elution profiles obtained in frontal and zonal chromatography of rat liver lactate dehydrogenase on 10-carboxydecylamino-Sepharose. **(a)** Advancing elution profiles obtained in frontal chromatography of enzyme (9 nM) on a 0.1 ml column in the absence of NADH ($\bigcirc$) and in the presence of 2 μM ($\bullet$), 3 μM ($\square$), 4 μM ($\blacksquare$), 5 μM ($\triangle$) and 6 μM ($\blacktriangle$) coenzyme. **(b)** Elution profiles obtained in zonal chromatography of enzyme (0.1 ml, 0.18 μM) on a 1.0 ml column in the presence of 8 μM ($\bigcirc$), 10 μM ($\bullet$), 12 μM ($\square$), 15 μM ($\blacksquare$) and 18 μM ($\triangle$) NADH. (Adapted from ref. 20 with permission).

The zonal method of quantitative affinity chromatography (1,2,5) enjoys greater popularity at present than the frontal method, presumably due to the fact that the latter technique is considered to employ too much solute. However, the difference between the amounts of solute required is not as great as might be expected at first sight. Although the volume of solution applied is much larger in a frontal than in a zonal experiment, the concentration can be considerably lower because of the absence of any dilution. This point is clearly evident from *Figure 1*, which compares the effect of NADH concentration on the advancing elution profile obtained in frontal affinity chromatography of rat liver lactate dehydrogenase on a 0.1 ml column (0.08 x 5.0 cm) of 10-carboxydecylamino-Sepharose with the corresponding profiles obtained by zonal chromatography on a 10-times larger column (0.50 x 1.28 cm) of the same affinity matrix (20).

The amount of lactate dehydrogenase used to generate the frontal profiles of *Figure 1a* (and hence to obtain two estimates of $\bar{V}_A$ in each case) ranged between 9 and 54 pmol (1 – 6 ml of 9 nM enzyme), compared with 18 pmol of enzyme (0.1 ml, 180 nM) for each zonal experiment (*Figure 1b*). It should also be noted that whereas the frontal patterns allow precise delineation of the median bisector of the boundary irrespective of the NADH concentration being studied, location of the equivalent position in the zonal profiles (*Figure 1b*) becomes increasingly difficult with decreasing ligand concentration. Finally, even though the zonal method does prove to be more economical in terms of solute the important point remains that results so obtained are uninterpretable unless certain simplifying ap-

155

proximations may be assumed. The above-mentioned study of the biospecific desorption by NADH of lactate dehydrogenase from 10-carboxydecylamino-Sepharose (20) is ample testimony to the fact that the frontal technique can be employed with relatively small amounts of partitioning solute.

3.2 Partition Equilibrium Experiments

An alternative to column chromatography is to conduct a series of partition equilibrium experiments in which the concentrations of solute in the liquid phase ($\bar{m}_A$) are determined directly for mixtures with known total concentration ($\bar{\bar{m}}_A$) of A (3,28). More specifically, mixtures containing known amounts of matrix and partitioning solute are prepared, (preferably by weight in order to define their composition most accurately), and then allowed to equilibrate at the temperature of interest until partition equilibrium has been established. A sample of the supernatant is then obtained by filtration (3) or centrifugation (28) of each mixture at the same temperature, and the weight concentration, $\bar{c}_A$, determined by, for example, enzymatic assay. The volumes $\bar{V}_A$ and V_A^* required for application of the appropriate expression from either *Table 1* or *Table 2* may then be obtained by means of the relationships $V_A^* \bar{\bar{c}}_A = \bar{V}_A \bar{c}_A$ and $\bar{\bar{c}}_A = w_A / V_A^*$, where w_A is the weight of the partitioning solute added to the slurry and V_A^* (the volume accessible to A) is determined from an experiment under conditions where there is no interaction of A with matrix sites or from experiments with matrix devoid of X groups (3,20). Any differences between the volumes of stationary (V_s) and liquid (V_0) phases in the various mixtures may be taken into account by employing the corresponding partition coefficient K_{av}^* (30) and the relationship $V_A^* = V_0 + K_{av}^* V_s$: the volume of the liquid phase may be deduced from an experiment with a solute that neither interacts with, nor penetrates into, the stationary phase.

A disadvantage inherent in partition experiments of the above design is the need to determine precisely the weight, and hence V_s (3) of affinity matrix present in each reaction mixture. Although this requirement poses no great problem if the matrix is available as an anhydrous solid, there is clearly a potential course of uncertainty if reliance must be placed on the reproducibility of aliquots derived from a concentrated slurry of affinity matrix. This difficulty can sometimes be obviated by the use of a recycling partition equilibrium technique (18,31) in which the liquid phase of a stirred slurry of affinity matrix and partitioning solute is analysed spectrophotometrically by means of a flow-cell placed in the line returning the liquid phase to the slurry (*Figure 2*). A number of partition experiments may therefore be performed with the same sample of matrix material by making a series of additions of concentrated solute to the slurry and noting the corresponding value of $\bar{c}_A$ after each addition. This recycling partition technique has the additional advantage that attainment of partition equilibrium is recognised by constancy of the absorbance of the continously monitored liquid phase.

3.3 The Problem of Defining the Free Ligand Concentration

The quantitative expressions for affinity chromatography (*Tables 1* and *2*) have been written in terms of m_S, the concentration of free ligand, which may only be

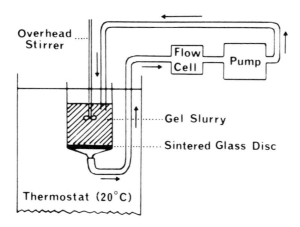

Figure 2. Schematic representation of the recycling partition equilibrium system. (Reproduced from ref.31 with permission).

equated with $\bar{m}_S$, the total ligand concentration in a prepared solute-ligand mixture if S does not interact with A, or if $\bar{m}_S \gg \bar{m}_A$. Initially, $\bar{m}_S$ was considered (3) to be the experimental quantity most readily available, and accordingly the quantitative analysis of univalent solutes was developed in terms of plots with $\bar{m}_S$ rather than m_S as the ligand concentration. However, since the use of $\bar{m}_S$ leads to curvilinearity of the suggested plots in cases 1 and 3, an experimenter would often be faced with the undesirable task of estimating limiting tangents (as $1/\bar{m}_S \rightarrow 0$) to the double-reciprocal plots. Furthermore, in the light of subsequent theoretical consideration (17 – 19), there is the distinct possibility that the curvilinearity may be reflecting solute multivalency, whereupon the limiting tangent approach is incorrect. The concept of using analyses based on $\bar{m}_S$ should therefore be discarded, but this raises the problem of defining the value of m_S in a solute-ligand mixture.

In affinity chromatography studies of protein-ligand interactions the concentration of protein ($\bar{m}_A$) is typically less than (say) 20 μM (1 mg/ml for a protein with a mol. wt. of 50 000). Accordingly, for studies requiring total ligand concentrations in excess of millimolar there is no problem in meeting the requirement that $\bar{m}_S \gg \bar{m}_A$ and therefore no error in the use of $\bar{m}_S$ as the estimate of m_S in the quantitative expressions. (*Tables 1* and *2*). However, if $\bar{m}_S$ and $\bar{m}_A$ are of comparable magnitudes, or indeed if $\bar{m}_S < \bar{m}_A$, there is clearly a need to distinguish between $\bar{m}_S$ and m_S; and to use the latter for the quantitative evaluation of the affinity chromatographic system.

Although it is possible in principle to employ iterative procedures based on parameters ($\bar{\bar{m}}_X$, k_{AX}, k_{AS}, etc.) obtained with $\bar{m}_S$ as an initial estimate of m_S (6), by far the simplest approach is to define m_S by exhaustive dialysis of the solute (protein) against a ligand solution of known concentration, and to use the diffusate for pre-equilibration of the affinity matrix. The consequent inability to know the magnitude of $\bar{m}_S$ is unimportant, the only ligand concentration required for application of the quantitative expressions from *Tables 1* and *2* being

m_S, which is defined unambiguously by the dialysis step. Because of the relatively high ionic strength (to minimise non-specific interactions) and low solute concentration employed in affinity chromatography, there should be effectively no Donnan correction even for charged ligands.

4. EVALUATION OF RESULTS FOR UNIVALENT SOLUTES

Any of the above designs of experiments yield values of $(\bar{m}_A, \bar{V}_A)$ from which it is necessary to determine four parameters, namely the two operative equilibrium constants, $\bar{\bar{m}}_X$ (the total concentration of accessible matrix sites) and V_A^* (the volume accessible to A). The feasibility of so doing is illustrated by considering results (3) for the glucose-dependent elution of lysozyme from Sephadex G-100, an example of case 3 and hence of the more difficult situation in which V_A^* is not measurable directly because of the interaction between A and matrix sites.

4.1 Complete Analysis of a Competitive Affinity System

Results of partition equilibrium experiments with the Sephadex G-100-lysozyme-glucose system (3) are presented in *Figure 3a*, which refers to experiments with $V_s = 4.44$ ml and $V_o = 7.80$ ml. Since the total concentration of enzyme in the liquid phase ($\bar{m}_A$) in these experiments (8.3 μM) was infinitesimal compared with that ($\bar{m}_S$) of glucose (0.17 – 0.84 M), m_S has been equated with $\bar{m}_S$. The first point to note about *Figure 3a* is the linearity of the double-reciprocal plot, which

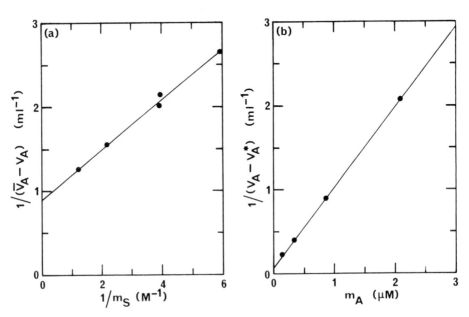

Figure 3. Graphical evaluation of equilibrium constants for the interactions of lysozyme with Sephadex G-100 (K_{AX}) and glucose (K_{AS}) by quantitative affinity chromatography. (a) Effect of glucose concentration on the elution volume ($\bar{V}_A$) of lysozyme in experiments with 8.3 μM enzyme. (b) Concentration dependence of the elution volume (V_A) for lysozyme in the absence of glucose. (Adapted from ref.3. with permission).

indicates that the system conforms with the concept that lysozyme is a univalent solute: this finding is reasonable in the sense that glucose, like N-acetyl-glucosamine (3), is presumably competing with the polysaccharide matrix for the single active site of the enzyme molecule. Linear regression analysis of the results shown in *Figure 3a* yields values of 0.30 for $(1 + K_{AX}\bar{m}_A)^2/V_A^*K_{AS}K_{AX}\bar{\bar{m}}_X$ and 0.90 for $(1 + K_{AX}\bar{m}_A)/V_A^*K_{AX}\bar{\bar{m}}_X$, these being obtained from the slope and intercept, respectively. Since $\bar{V}_A \rightarrow V_A^*$ as $1/m_S \rightarrow 0$, the intercept also corresponds to $1/(\bar{V}_A - V_A^*)$. A combination of the consequent value of 1.11 ml for $(V_A - V_A^*)$ with that of 11.73 ml for V_A at this particular $\bar{m}_A$ (8.3 μM) gives $V_A^* = 10.62$ ml. The format of *Figure 3b*, which depicts the dependence of V_A upon enzyme concentration in the absence of glucose, is suggested by combination of the two expressions for the ordinate intercept of *Figure 3a*, which leads to the relationship

$$\frac{1}{(V_A - V_A^*)} = \frac{1}{V_A^*K_{AX}\bar{\bar{m}}_X} + \frac{\bar{m}_A}{V_A^*\bar{\bar{m}}_X} \qquad \text{Equation 3}$$

From the slope (9.46 x 10⁴) and intercept (0.079) of the semi-reciprocal plot in *Figure 3b*, values of 1.0 μM and 1.2 x 10⁶ M⁻¹ are obtained for $\bar{\bar{m}}_X$ and K_{AX}, respectively. The remaining parameter, K_{AS}, is now calculated to be 30 M⁻¹ from the slope of *Figure 3a*.

4.2 Analysis of Ligand-retarded Affinity Systems

Thus far, consideration has been restricted to a situation where the partitioning solute interacts directly with matrix sites (case 3; also applicable to case 4). Attention is now turned briefly to situations where the interaction of solute with matrix sites is via ternary complex formation. Of the two possible situations (cases 1 and 2), that in which the soluble AS complex interacts with matrix-bound X sites (case 1) is more likely to be encountered in quantitative affinity chromatography, since the logical response to the other possibility, viz., the requirement that S must first interact with X, would be to employ a matrix with S as a covalently bound part of an immobilised reactant XS. In either case there is clearly no information, apart from the value of V_A^*, to be gleaned from experiments in the absence of ligand, and hence $\bar{m}_X$ must be obtained from experiments with solute-ligand mixtures.

Inspection of the expressions for cases 1 and 2 in *Table 1* shows that the intercept/slope ratio of the plot of $1/(\bar{V}_A - V_A^*)$ *versus* $1/m_S$ yields the product $K_B(1 + K_T\bar{m}_A)$, where K_B, the binary equilibrium constant, is either K_{AS} (case 1) or K_{XS} (case 2); and K_T, the ternary equilibrium constant, is either K_{ASX} (case 1) or K_{XSA} (case 2). Consequently, determination of the same product from a second series of experiments with a different value of $\bar{m}_A$ in principle provides the second simultaneous equation required for evaluating the two parameters K_B and K_T: $\bar{m}_X$ then follows. Alternatively, by determining the dependence of $\bar{V}_A$ upon solute concentration ($\bar{m}_A$) in a series of experiments with sufficiently high ligand concentration that $K_B m_S$ is very large, a plot of $1/(V_A - V_A^*)$ *versus* m_A may be used to define $\bar{m}_X$ and K_T from the slope and the intercept. Indeed, the expression for ligand-retarded affinity chromatography in the presence of a saturating concentration of ligand becomes formally identical with that used (equation 3)

159

for ligand-facilitated systems in the absence of ligand, the only differences being the substitution of a ternary equilibrium constant (K_{ASX} or K_{XSA}) for the binary equilibrium constant K_{AX}; and of $\bar{V}_A$ for V_A. No experimental studies have yet been reported which provide evidence of the relative merits of these two procedures.

4.3 Approximate Analysis of Affinity Chromatographic Results

The above analysis of the Sepharose-lysozyme-glucose system has shown that characterisation of this competitive situation (K_{AX}, K_{AS}) requires a series of experiments with different solute concentrations (*Figure 3b*) as well as the series (at constant $\bar{m}_A$) in which m_S is varied (*Figure 3a*), thereby reinforcing the earlier statement that two series of experiments are invariably required in order to provide a complete characterisation of an affinity system. However, in many studies the primary objective is only the determination of the equilibrium constant for the solute-ligand interaction, and it is therefore logical to enquire whether fewer experiments might suffice for such partial characterisation of an affinity chromatographic system.

Inspection of the expression in *Table 1* shows that the ratios of the slope to the intercept of the suggested plots for analysis of the dependence of $\bar{V}_A$ upon ligand concentration all contain $\bar{m}_A$, the total solute concentration in the plateau region of a frontal experiment. For example, in the situation under consideration (case 3), the slope/intercept yields $(1 + K_{AX}\bar{m}_A)/K_{AS}$. However, provided that $K_{AX}\bar{m}_A \ll 1$, this ratio provides a direct measure of K_{AS}, and hence it is possible in principle to select a solute concentration that is sufficiently low for the approximation to be made (3,6). The problem with this approximate approach is the uncertainty about the magnitude of K_{AX}, which may be much greater than K_{AS} even in instances where the immobilised reactant and the ligand are essentially identical entities (18). For example, use of the approximation with the results presented in *Figure 3a* would have led to the conclusion that $K_{AS} = 3$ M^{-1} for the lysozyme-glucose interaction, which, if used as an estimate of K_{AX} for the interaction between enzyme and polyanhydroglucose matrix, would seemingly provide retrospective justification for neglecting the $K_{AX}\bar{m}_A$ term in these experiments with $\bar{m}_A = 8.3$ μM. However, from the studies with lysozyme alone (*Figure 3b*) it is clear that the magnitude of K_{AX} bears no resemblance to that of K_{AS}; and that $K_{AX}\bar{m}_A \cong 10$, which is the source of the 10-fold underestimation of K_{AS} by the approximate method.

Neglect of the $K_{AX}\bar{m}_A$ term is implicit in the analysis of frontal chromatographic results by the method of Kasai and Ishii (4). It also forms the basis of the method for analysing zonal chromatographic results (2,3,5), where, in the absence of any value for $\bar{m}_A$, the only course of action open is to assume that the approximation $K_{AX}\bar{m}_A \ll 1$ pertains. Certainly there are many systems for which the approximate method has been validated by agreement between the K_{AS} so obtained and that found by other methods. However, it is clearly preferable to perform the complete analysis, whereby the validity of the result is not conditional upon its verification by other methods.

5. ANALYSIS OF RESULTS FOR MULTIVALENT SOLUTES

In only four quantitative affinity chromatographic studies has account been taken of the multiple interaction of solute species with matrix sites; and in one of those (14) an approximate analysis (17) of zonal chromatographic results was used to interpret the phosphorylcholine-dependent elution of immunoglobulin A from Sepharose to which phosphorylcholine had been covalently attached as the immobilised reactant X. Other investigations have considered the phosphate-dependent elution of aldolase from phosphocellulose (18) and from rabbit muscle myofibrils (28), and the NADH-dependent elution of lactate dehydrogenase from 10-carboxydecylamino-Sepharose (32). All four systems are examples of case 3, involving competition between S and X for solute sites.

5.1 The Total Concentration of Accessible Sites

An essential part of quantitative affinity chromatographic investigations involving multivalent solutes, as in studies with their univalent counterparts, is the determination of the total concentration of accessible matrix sites. At this stage procedures for quantitative assessment of $\bar{\bar{m}}_X$ have been developed and tested for systems in which A interacts directly with X (specifically case 3 but also applicable to case 4). However, due to the lack of any reported studies of multivalent solute systems exhibiting ligand-retarded elution (cases 1 and 2), there is again no experimentally proven protocol for such systems. Accordingly attention is focussed mainly on methods for measurement of $\bar{\bar{m}}_X$ for systems conforming to case 3 and case 4, after which their possible adaptation to analysis of the other situations is considered. As with the corresponding univalent solute systems, $\bar{\bar{m}}_X$ and the solute-matrix equilibrium constant, k_{AX}, are first determined from the extent of interaction of solute with affinity matrix in the absence of ligand. Three procedures have been suggested for determination of the effective total concentration of accessible matrix sites, $\bar{\bar{m}}_X$, from measurement of the amount of solute bound as a function of solute concentration in the liquid phase, m_A, which is equivalent to m_A since A is the only soluble form of solute in the absence of ligand.

5.2 Extrapolation of Results to Infinite Solute Concentration

The inital approach (18) to the problem of defining $\bar{\bar{m}}_X$ was to extrapolate results to infinite solute concentration, the rationale of this method being that as $\bar{m}_A \rightarrow \infty$, all accessible matrix sites become saturated with singly bound solute molecules (27). Thus, in recycling partition equilibrium studies of the phosphocellulose-aldolase system the ordinate intercept obtained from a plot of the reciprocal amount of aldolase bound, $1/V_A^*(\bar{\bar{m}}_A - \bar{m}_A)$, against $1/\bar{m}_A$ has been used (18) as a means of determining the amount ($V_A^* \bar{\bar{m}}_X$) of matrix sites. Essentially the same procedure, but with the amount of enzyme bound expressed as a binding function, $V_A^*(\bar{\bar{m}}_A - \bar{m}_A)/w_X$ where w_X is the weight of affinity matrix, has also been used in a partition equilibrium study of the interaction between aldolase and muscle myofibrils (28). Since, as noted above, $\bar{V}_A \bar{m}_A = V_A^* \bar{\bar{m}}_A$, the logical plot of frontal chromatographic results would be $1/\bar{m}_A(\bar{V}_A - V_A^*)$ *versus*

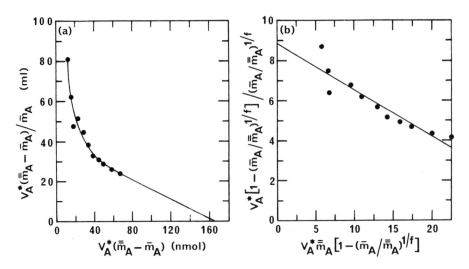

Figure 4. Analysis of partition equilibrium results (Table 1 of ref.18) for the interaction of aldolase with phosphocellulose. (a) Evaluation of the molar amount $(V_A^* \bar{\bar{m}}_X)$ of matrix sites for aldolase in 0.783 g of phosphocellulose by extrapolation of results to infinite enzyme concentration. (b) Plot of the results in accordance with the appropriate form of equation 4 for the simultaneous determination of k_{AX} and $V_A^* \bar{\bar{m}}_X$ on the basis that aldolase is tetravalent $(f = 4)$.

$1/\bar{m}_A$, the ordinate intercept of which would equal $1/V_A^* \bar{\bar{m}}_X$.

As noted in the original publication of this approach (18), the extrapolation of such double-reciprocal plots is somewhat uncertain, a point made much more dramatically by plotting the same results in Scatchard (33) format (*Figure 4a*): the abscissa intercept of *Figure 4a* has been chosen to coincide with the earlier inference (18) that 167 nmol of aldolase was required to saturate the accessible phosphocellulose sites. Despite the obvious uncertainty associated with such extrapolations, the extrapolated value does provide an estimate of $\bar{\bar{m}}_X$, whereupon k_{AX} may be obtained from each experimental measurement, $(\bar{m}_A, \bar{\bar{m}}_A)$ or $(\bar{m}_A, V_A^*/\bar{V}_A)$, as the only remaining parameter of unknown magnitude in the quantitative expression describing the binding of tetravalent enzyme (*Table 2*, case 3 with $f = 4$ and $m_S = 0$): a selection of such results from ref.18 is summarised in the first part of *Table 3*. Furthermore the value of $\bar{\bar{m}}_X$ is amenable to refinement by ascertaining improved constancy of the values of k_{AX} so deduced.

5.3 Simultaneous Evaluation of k_{AX} and $\bar{\bar{m}}_X$ by Graphical Means

Although the above procedure for defining $\bar{\bar{m}}_X$ has been employed in two investigations (18,28), it is clearly preferable to avoid such extrapolations of results. Recent reconsideration of this problem (22) has led to rearrangement of case 3 to the form:

$$\frac{[1 - (V_A^*/\bar{V}_A)^{1/f}]}{(V_A^*/\bar{V}_A)^{1/f}} = \frac{k_{AX}\bar{\bar{m}}_X}{1 + k_{AS}m_S} - \frac{fk_{AX}[1 - (V_A^*/\bar{V}_A)^{1/f}](\bar{V}_A/V_A^*)\bar{m}_A}{1 + k_{AS}m_S} \quad \text{Equation 4}$$

Table 3. Evaluation of Site-binding Constants for the Aldolase-Phosphocellulose and Aldolase-Phosphate Interactions by Affinity Chromatography[a].

m_S (mM)	$\bar{\bar{m}}_A$ (μM)	$\bar{m}_A$ (μM)	$\bar{\bar{m}}_X$ (μM)	$10^{-4}\,k_{AX}{}^a$ (M^{-1})	$k_{AS}{}^b$ (M^{-1})
–	1.33	0.25	11.60	5.32	–
–	2.08	0.46	11.36	5.24	–
–	3.15	0.89	11.02	4.89	–
–	4.27	1.44	10.67	4.72	–
–	5.83	2.28	10.18	4.98	–
–	6.78	2.80	9.88	5.48	–
1.00	0.113	0.069	3.09		180
1.00	0.224	0.143	3.07		280
1.00	0.335	0.221	3.06		360
1.00	0.662	0.443	3.02		340
1.00	0.769	0.509	3.01		270
1.00	0.875	0.593	3.00	5.1^c	330
5.00	0.080	0.055	3.58		170
5.00	0.160	0.125	3.57		370
5.00	0.239	0.191	3.57		420
5.00	0.319	0.261	3.56		490
5.00	0.397	0.322	3.55		460
5.00	0.554	0.456	3.54		500

[a]Results are from Tables 1 and 2 of ref. 18.
[b]Calculated on the basis of the expression for case 3 in *Table 2* (see text).
[c]The mean value of k_{AX} from the experiments with $m_S = 0$.

from which it is evident that for an assigned value of the solute valence f, $\bar{\bar{m}}_X$ may in principle be obtained from the ratio of the slope to the intercept of a plot of $[1 - (V_A^*/\bar{V}_A)^{1/f}]/(V_A^*/\bar{V}_A)^{1/f}$ *versus* $[1 - (V_A^*/\bar{V}_A)^{1/f}](\bar{V}_A/V_A^*)\bar{m}_A$ for results obtained with the same ligand concentration m_S and a range of $\bar{m}_A$ values. Application of this expression to the results of the partition equilibrium experiments on the aldolase-phosphocellulose system (18) in the absence of ligand ($m_S = 0$) is most readily accomplished by noting that $(V_A^*/\bar{V}_A) = (\bar{m}_A/\bar{\bar{m}}_A)$. *Figure 4b* shows the relevant plot of the aldolase-phosphocellulose results in these terms, the values of $V_A^*\bar{\bar{m}}_X$ and k_{AX} obtained by least square calculations being 154 ($\pm$18) nmol and 5.8 ($\pm$0.8) x 10^4 M^{-1}, respectively. It should be noted that these findings vindicate the earlier treatment (18) of the results based on the extrapolation procedure in as much as the magnitudes of 167 nmol for $V_A^*\bar{\bar{m}}_X$ (*Figure 4a*) and 5.1 x 10^4 M^{-1} for k_{AX} (*Table 3*) are encompassed by the present analysis.

A legitimate objection to this second method of analysis is, of course, its reliance upon the value selected for f, the solute valence. Although in the present instance this requirement for knowledge of the number of active sites on the tetrameric aldolase molecule poses no problem, there are likely to be many systems for which the magnitude of f would also have to emanate from analysis of the affinity chromatography results. In this regard the extrapolation method for obtaining $\bar{\bar{m}}_X$, despite its inherent uncertainty, is independent of the value of f, and hence the magnitude of the valence may in principle then be evaluated by analysing the experimental results (*Table 3*) for a range of possible f values, and selecting the appropriate valence on the basis of constancy of k_{AX}. However, because of the uncertainty in the required extrapolation, that type of approach has already been suggested above as the means of refining the value of $\bar{\bar{m}}_X$ for a given f. A possible course of action to overcome this dilemma is the joint use of both approaches.

Repetition of the analysis of the aldolase-phosphocellulose results on the basis of Equation 4 with $f = 2$ yields values of 124 (± 15) nmol and 1.8 (± 0.5) x 10^5 M^{-1} for the apparent values of $V_A^* \bar{\bar{m}}_X$ and k_{AX}, respectively. This analysis is, however, less acceptable from the viewpoint that the magnitude of $V_A^* \bar{\bar{m}}_X$ so calculated is not readily reconciled with the f-independent value obtained by the extrapolation procedure (*Figure 4a*). Furthermore, the relative error associated with the estimate of k_{AX} is 30%, or more than 2-fold greater than that deduced from *Figure 4b* on the basis that $f = 4$, a finding that also favours the higher value for the solute valence. Indeed, evaluation of parameters by minimisation of standard errors forms the basis of a third procedure for determining $\bar{\bar{m}}_X$ and k_{AX}.

5.4 Evaluation of $\bar{\bar{m}}_X$ and k_{AX} by Standard Error Minimisation

A third procedure for determining $\bar{\bar{m}}_X$ has been suggested (32) in relation to analysis of frontal chromatographic results (20) on the interaction of lactate dehydrogenase with 10-carboxydecylamino-Sepharose. For this system the range of enzyme concentrations examined (20) proved insufficient to allow precise delineation of the ordinate intercept of the suggested (18) double-reciprocal plot to obtain $\bar{\bar{m}}_X$ (*Figure 5a*). However, the appropriate quantitative relationship in *Table 2* (case 3, $m_S = 0$) may be used to determine apparent magnitudes of k_{AX} from each of the six experimental points for a range of $\bar{\bar{m}}_X$ values, and the appropriate value taken as that giving the smallest standard error in the mean value of k_{AX} so determined. From *Figure 5b*, which shows the dependence on $\bar{\bar{m}}_X$ of the relative error (twice the S.E.M. expressed as a percentage of the mean) in k_{AX} calculated for $f = 4$, the results are best described by values of 91 μM and 1.8 x 10^4 M^{-1} for $\bar{\bar{m}}_X$ and k_{AX}, respectively. With $f = 2$ the value of $\bar{\bar{m}}_X$ for minimal relative error in k_{AX} decreased to 46 μM, but the extent of the relative error in k_{AX} (6.8%) was greater, thereby signifying better conformity of the results with the concept of the partitioning enzyme acting as a tetravalent species ($f = 4$). Thus the procedure based on error minimisation also offers considerable potential in the quantitative appraisal of affinity chromatography results for the most appropriate value of $\bar{\bar{m}}_X$.

5.5 Systems Exhibiting Ligand-retarded Elution of Solute

One possible method for evaluating $\bar{\bar{m}}_X$ for such systems is to employ essentially the second procedure described above to results obtained for a range of m_A values in the presence of a fixed concentration, m_S, of ligand, the analogous rearrangement of the quantitative expressions from *Table 2* being:

$$\frac{[1 - (V_A^*/\bar{V}_A)^{1/f}]}{(V_A^*/\bar{V}_A)^{1/f}} = \frac{k_B m_S k_T \bar{\bar{m}}_X}{1 + k_B m_S} - \frac{f k_B m_S k_T [1 - (V_A^*/\bar{V}_A)^{1/f}] (\bar{V}_A/V_A^*) \bar{m}_A}{1 + k_B m_S}$$

Equation 5

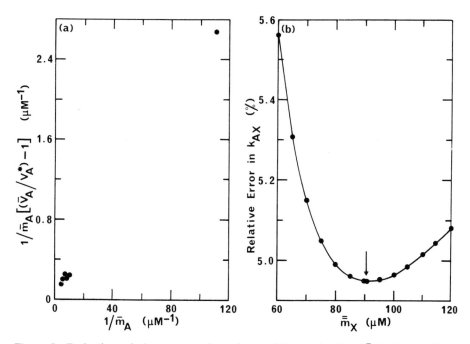

Figure 5. Evaluation of the concentration of accessible matrix sites ($\bar{\bar{m}}_X$) from a frontal chromatographic study (20) of the interaction between rat liver lactate dehydrogenase and 10-carboxydecylamino-Sepharose. **(a)** The suggested (18) double-reciprocal plot of the results (reported in Figure 2 of ref.20) for determining $\bar{\bar{m}}_X$ as the reciprocal of the ordinate intercept. **(b)** Identification of $\bar{\bar{m}}_X$ from the same results *via* the procedure based on minimisation of the relative error in k_{AX} values calculated by assigning magnitudes to $\bar{\bar{m}}_X$ (32).

where k_B and k_T are the respective equilibrium constants for the formation of binary and ternary complexes, as in the discussion of the analogous cases for univalent solutes. Irrespective of the fixed value of m_S selected, the ratio of the slope to the intercept of a plot of $[1-(V_A^*/\bar{V}_A)^{1/f}]/(V_A^*/\bar{V}_A)^{1/f}$ *versus* $[1-(V_A^*/\bar{V}_A)^{1/f}](\bar{V}_A/V_A^*)\bar{m}_A$ is suggested as one possible means of defining $\bar{\bar{m}}_X$. Furthermore, if the ligand concentration is sufficiently high to allow the approximation that $k_B m_S \gg 1$, the value of the intrinsic association constant for ternary complex formation may be obtained from the slope. As noted in relation to the univalent case, the proviso ($k_B m_S \gg 1$) renders the expression for ligand-retarded affinity chromatography in the presence of ligand formally identical with that for ligand-facilitated elution in the absence of ligand (*Table 2*), the only difference being the substitution of a ternary equilibrium constant (k_{ASX} or k_{XSA}) for the binary equilibrium constant, k_{AX}. Consequently, all three of the above procedures are in principle applicable to results obtained for ligand-retarded affinity systems in the presence of a saturating concentration of ligand. As mentioned previously, no experimental studies have yet been performed to provide a test of this conclusion.

5.6 Evaluation of the Second Equilibrium Constant

Of the four parameters for which magnitudes are to be deduced from quantitative affinity chromatographic studies (f, $\bar{\bar{m}}_X$, and two equilibrium constants), the solute valence and one equilibrium constant are also evaluated during the process of evaluating $\bar{\bar{m}}_X$. Consequently, there only remains the problem of obtaining a magnitude for the other intrinsic assocation constant, a relatively simple task. Irrespective of the situation (cases $1-4$), the experimental information required is the quantity $(V_A^*/\bar{V}_A)$ or $(\bar{m}_A/\bar{\bar{m}}_A)$ for reaction mixtures containing a range of ligand concentrations, m_S. In frontal chromatography, for example, this could entail the determination of $\bar{V}_A$ as a function of m_S in experiments with fixed $\bar{m}_A$ (20,32), whereas in the recycling partition equilibrium procedure (18) it is more likely that a range of solute concentrations would be examined at each of several ligand concentrations. *Table 3* concludes with a selection of such results for the aldolase-phosphocellulose system in the presence of phosphate (18), and the value of k_{AS} obtained by application of the quantitative expression (case 3) from *Table 2* to each experimental point. On the basis of *Table 3* it may therefore be concluded (18) that the phosphate-dependent elution of rabbit muscle aldolase from phosphocellulose reflects competition between phosphate and matrix sites for four equivalent and independent sites on aldolase, the magnitudes of k_{AX} and k_{AS} being 5.1×10^4 M^{-1} and 350 M^{-1}, respectively, at pH 7.4, I $= 0.15$.

6. SIGNIFICANCE OF EQUILIBRIUM CONSTANTS FOR MATRIX INTERACTIONS

With regard to the above study of the phosphocellulose-aldolase-phosphate systems it is noted that although the magnitude of k_{AS} describes the interaction of aldolase with phosphate in conventional solution terms, the value of the enzyme-matrix association constant is governed to some extent by the volume frame of reference chosen to define $\bar{\bar{m}}_X$, i.e., the volume accessible to solute. Despite its somewhat empirical nature, k_{AX} still suffices from the thermodynamic viewpoint of determining the distribution of enzyme between soluble and immobilised states in any reaction mixture for which the matrix concentration ($\bar{\bar{m}}_X$) and total solute concentration ($\bar{\bar{m}}_A$) are defined on the same volume scale as that to which k_{AX} refers. This is an important consideration in studies of naturally occurring affinity chromatographic systems such as the metabolite-dependent distribution of aldolase in muscle (28). However, if the aim of a quantitative affinity chromatographic study is to determine the magnitude of the binding constant for a solute-ligand interaction, k_{AX} for the binding of solute to ligand immobilised on matrix should clearly not be regarded as an estimate of the equilibrium constant for the corresponding interaction between solute and ligand in solution; not only because of the dependence of the magnitude of k_{AX} upon some empirically closed volume scale for the definition of $\bar{\bar{m}}_X$ and $\bar{\bar{m}}_A$, but also because of fundamental differences between the two interactions arising from steric effects and the consequences of the chemical modification associated with immobilisation of the ligand. Consequently, the equilibrium constant for the solute-ligand interaction of interest should be obtained as k_{AS}, possibly from experiments in which soluble and immobilised forms of the same ligand compete for solute: this was the procedure adopted by Eilat and Chaiken (14) to investigate the interaction between immunoglobulin A and phosphorylcholine.

7. CONCLUDING REMARKS

The past decade has seen the emergence of quantitative affinity chromatography as a method for studying interactions between dissimilar reactants. The method, which is still in its relative infancy, has the undeniable attraction that the specificity of the interaction with the immobilised reactant may in some circumstances make possible the quantitative assessment of a reversible reaction occurring within a biological milieu, a factor demonstrated by the use of Sepharose-oxamate to study the interactions between NADH and the various lactate dehydrogenase isoenzymes in a crude tissue extract (6). Another advantage of affinity chromatography would seem to be the lack of any restriction with regard to the range of magnitudes of equilibrium constants that may be measured by the technique. From the two examples considered in detail herein, it is evident that quantitative affinity chromatography is eminently suited to the characterisation of relatively weak interactions ($k_{AS} < 1000$ M^{-1}), which prove troublesome in methods such as equilibrium dialysis because of the difficulties encountered in measuring the difference between total and free ligand concentrations. Quantitative affinity chromatography should, however, also prove useful for evaluating interactions governed by equilibrium constants at the other end of the spectrum of magnitudes: those interactions for which k_{AS} is so large that the range of ligand concentrations required for their study is too low for experimental detection, let alone quantitative measurement, in a reaction mixture. By predialysing the solute exhaustively against a series of ligand solutions, the concentrations of which have been defined through their preparation by weight dilution of a solution with an experimentally measurable concentration, it should be possible to define $\bar{V}_A$ as a function of m_S; and thereby obtain k_{AS} for the interaction.

Although the theory on which this review is based undoubtedly suffices for the study of a large number of interactions, the coverage is by no means comprehensive. For example, the possible interaction of both A and liganded A with the matrix is a situation (k_{AX} and k_{ASX} both operative) that is not considered in the present treatment; but it is one for which theory, albeit more complicated, is available (19). However, the area in which the current treatments of the quantitative affinity of multivalent solutes (18,19) become most restrictive is the assumption that a single intrinsic association constant describes all solute-matrix interactions of a particular type. It is not difficult to envisage situations in which successive interactions of a solute with a matrix may be characterised by either enhanced or diminished site-binding constants as a result of the steric requirements imposed by the placement of the immobilised reactant groups X. Another restrictive aspect of the present treatments, already commented upon, is the assumed identity of the gel partitioning characteristics of solute and all solute-ligand complexes. Furthermore, no account has been taken of the kinetics (chemical and mass transfer) of the partition process. In this regard, a more general theory of quantitative affinity chromatography (34) has shown that the constraints on the magnitudes of rate constants necessitated by the assumed attainment of partition equilibrium in the simplified treatment are likely to be of no concern in conventional column chromatographic studies; but could possibly render invalid the application of the present expressions to results obtained using the much faster flow-rates associated with h.p.l.c. As noted by Chaiken (35), the potential use of affinity chromatography for quantitative studies of ligand binding still clearly requires further exploration, from theoretical as well as experimental viewpoints. However, from the studies so far conducted it would appear that quantitative affinity chromatography may well prove to be one of the most versatile methods available for studying interactions between ligands and macromolecules.

8. REFERENCES

1. Andrews,P., Kitchen,B.J. and Winzor,D.J. (1973) *Biochem. J.*, **135**, 897.
2. Dunn,B.M. and Chaiken,I.M. (1974) *Proc. Natl. Acad. Sci. USA*, **71**, 2382.
3. Nichol,L.W., Ogston,A.G., Winzor,D.J. and Sawyer,W.H. (1974) *Biochem. J.*, **143**, 435
4. Kasai,K. and Ishii,S. (1975) *J. Biochem. (Tokyo)*, **77**, 261.
5. Dunn,B.M. and Chaiken,I.M. (1975) *Biochemistry (Wash.)*, **14**, 2343.
6. Brinkworth,R.I., Masters,C.J. and Winzor,D.J. (1975) *Biochem. J.*, **151**, 631.
7. Chaiken,I.M. and Taylor,H.C. (1976) *J. Biol. Chem.*, **251**, 2044.
8. Taylor,H.C. and Chaiken,I.M. (1977) *J. Biol. Chem.*, **252**, 6991.
9. Kasai,K. and Ishii,S. (1978) *J. Biochem. (Tokyo)*, **84**, 1051.
10. Danner,J., Sommerville,J.E., Turner,J. and Dunn,B.M. (1979) *Biochemistry (Wash.)*, **18**, 3039.
11. Dunn,B.M. and Gilbert,W.A. (1979) *Arch. Biochem. Biophys.*, **198**, 533.
12. Oda,Y., Kasai,K. and Ishii,S. (1981) *J. Biochem. (Tokyo)*, **89**, 285.
13. Veronese,F.M., Bevilacqua,R. and Chaiken,I.M. (1979) *Mol. Pharmacol.*, **15**, 313.
14. Eilat,D. and Chaiken,I.M. (1979) *Biochemistry (Wash.)*, **18**, 790.
15. Kozutsumi,Y., Kawasaki,T. and Yamashina,I. (1981) *J. Biochem. (Toyko)*, **90**, 1799.
16. Kálmán,M. and Boross,L. (1982) *Biochim. Biophys. Acta*, **704**, 272.
17. Chaiken,I.M., Eilat,D. and McCormick,W.M. (1979) *Biochemistry (Wash.)*, **18**, 794.
18. Nichol,L.W., Ward,L.D. and Winzor,D.J. (1981) *Biochemistry (Wash.)*, **20**, 4856.
19. Winzor,D.J., Ward,L.D. and Nichol,L.W. (1982) *J. Theor. Biol.*, **98**, 171.
20. Kyprianou,P. and Yon,R.J. (1982) *Biochem. J.*, **207**, 549.
21. Winzor,D.J. (1981) in *Protein-Protein Interactions*, Frieden,C. and Nichol,L.W. (eds.), Wiley, New York, p.129.
22. Hogg,P.J. and Winzor,D.J. (1984) *Arch. Biochem. Biophys.*, in press.
23. Klotz,I.M. (1946) *Arch. Biochem.*, **9**, 109.
24. Nichol,L.W. and Winzor,D.J. (1981) in *Protein-Protein Interactions*, (Frieden,C. and Nichol, L.W. (eds.), Wiley, New York, p.337.
25. Flory,P.J. (1941) *J. Am. Chem. Soc.*, **63**, 3083.
26. Singer,S.J. (1965) in *The Proteins*, Vol. 3, Neurath,H. (ed.), Academic Press, New York, p. 269.
27. Calvert,P.D., Nichol,L.W. and Sawyer,W.H. (1979) *J. Theor. Biol.*, **80**, 233.
28. Kuter,M.R., Masters,C.J. and Winzor,D.J. (1983) *Arch. Biochem. Biophys.*, **225**, 384.
29. Longsworth,L.G. (1943) *J. Am. Chem. Soc.*, **65**, 1755.
30. Laurent,T.C. and Killander,J. (1964) *J. Chromatogr.*, **14**, 317.
31. Ford,C.L. and Winzor,D.J. (1981) *Anal. Biochem.*, **114**, 146.
32. Winzor,D.J. and Yon,R.J. (1984) *Biochem. J.*, **217**, 867.
33. Scatchard,G. (1949) *Ann. N.Y. Acad, Sci.*, **51**, 660.
34. Hethcote,H.W. and De Lisi,C. (1982) *J. Chromatogr.*, **248**, 183.
35. Chaiken,I.M. (1979) *Anal. Biochem.*, **97**, 1.

CHAPTER 7

Zonal Elution Quantitative Affinity Chromatography and Analysis of Molecular Interactions

D.M. ABERCROMBIE AND I.M. CHAIKEN

1. INTRODUCTION

Interactions of small molecules, macromolecules and multimolecular assemblies can be measured from the elution characteristics of these substances on affinity matrices containing immobilised interactants. In affinity chromatography, a common goal is to prepare an immobilised molecular derivative which retains the ability to bind biospecifically to a mobile substance, often a protein or other macromolecule (*Figure 1*). When such specificity is achieved, powerful separation methods are available for preparative isolation of the desired species from complex mixtures which may contain contaminants closely related in other properties often used as bases for purification, such as size and non-biospecific surface characteristics. As opposed to preparative isolation, affinity chromatography also can be a flexible analytical method, since the degree of chromatographic retardation of a mobile interactant on an immobilised interactant is a measure of the equilibrium binding constant for the affinity matrix interaction of the mobile component (1,2). Further, if, during zonal elution, molecules are included which affect affinity matrix binding by direct interaction with the eluting

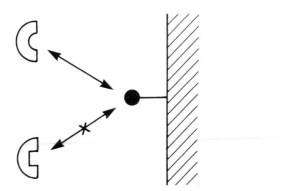

Figure 1. Schematic diagram showing theoretical discrimination of affinity matrix in recognising a desired mobile interactant by virtue of a specific binding surface. When such biospecificity is achieved with an affinity matrix, the latter can be used analytically to study quantitative features of the binding interaction between mobile and immobilised interactants and between these and mobile effectors.

mobile component (as by competition with the immobilised component), binding parameters for interaction of such effectors also can be measured. And, since, for a given affinity matrix, competitors and other effectors can be varied, as can the chemical nature of mobile interactant, binding specificity and dependence on solution conditions such as pH, buffer components and temperature can all be determined. Finally, when binding constants for the immobilised interactant can be compared with those for the soluble analogue, the relatedness of affinity matrix and solution interactions can be used in designing and evaluating affinity matrices as preparative tools by helping to define the degree of fidelity of an affinity chromatography matrix as a biospecific interactant.

The zonal elution approach of analytical affinity chromatography was developed essentially in parallel with the continuous elution/frontal analysis approach (see Chapter 6). Owing largely to its simplicity of experimental design, the zonal method has become widely applicable for biochemical analysis of biomolecular interactions when one of the molecular partners can be immobilised with retention of interaction properties. Since the method can be used to analyse very small amounts of the mobile interactant, it has important applicability as a micromethod for biochemical analysis of a vast array of biologically active molecules being discovered and isolated but available in only small amounts (for example, as immunoreactive or radiolabelled species).

This contribution describes the basic experimental principles and techniques of zonal elution analytical affinity chromatography. Features of affinity matrix and mobile interactant preparations particularly applicable for analytical (as opposed to preparative) experiments are discussed. Finally, representative examples are given showing the types of result and information which have been obtained for several protein-ligand and protein-protein interacting systems.

2. ZONAL ELUTION IN ANALYTICAL AFFINITY CHROMATOGRAPHY

The basic procedure of zonal elution affinity chromatography of a mobile interacting molecule on a matrix of immobilised interactant is:

(i)   to apply a zone of small volume containing the mobile component onto a column equilibrated with appropriate buffer with or without effector,

(ii) to continue eluting with equilibration buffer (+ / − effector), and

(iii) to determine the elution profile of the initially applied mobile component.

The types of interactions which can occur during affinity chromatography when the effector is a competitive ligand are shown in *Figure 2*; and an example of the type of results expected is shown in *Figure 3*.

The following may be considered as general methodological features for zonal elution affinity chromatography.

(i)   Zonal elution is usually performed with small bed columns, typically of volumes of 1 − 10 ml with narrow internal diameters (7 − 10 mm).

(ii) After packing the matrix, the column is equilibrated with the buffer to be used throughout the chromatographic elution. The choice of running columns at a particular temperature depends largely on the stability of the

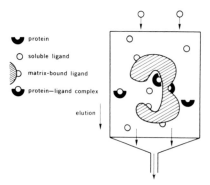

Figure 2. Schematic diagram of competitive zonal elution affinity chromatography, depicting the interactions of mobile and immobilised interactants. Here, the elution buffer contains a soluble ligand that competes with matrix-bound ligand for binding to the same active site of the mobile interactant (depicted here as a protein). Adapted from (3).

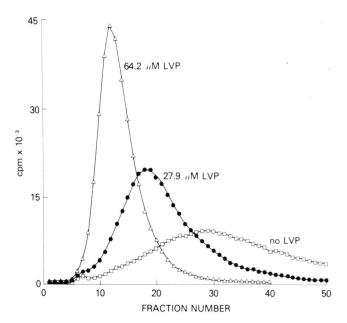

Figure 3. A representative series of competitive zonal elution profiles, taken from neurophysin/peptide studies, showing the chromatographic behaviour of zones of protein eluted on immobilised ligand with increasing amounts of soluble competing ligand added to the elution buffer. Zones (each of 100 μl applied to a bed volume of about 2 ml) of ^{125}I-labelled bovine neurophysin II were eluted from Met-Tyr-Phe-aminobutyl-agarose with 0.4 M ammonium acetate, pH 5.7, containing the competing ligand, lysine-vasopressin, at varying concentrations (μM): 0 ($\square$); 27.9 ($\bullet$); 64.2 ($\triangle$). Increasing the concentration of soluble competitor results in proportionately lower elution volumes of the labelled protein. (Adapted from ref. 4).

molecular interactants and the nature of the non-covalent forces between mobile and immobilised components.

(iii) Under optimal conditions several mechanical column parameters (i.e., flow-rate, V_O (the elution volume of a non-interacting, penetrable molecule

171

of size similar to the mobile interactant) and V_m (the void or gel-excluded volume) are kept constant throughout a series of zonal elution experiments. This enables comparisons of results among many chromatographic analyses and makes it unnecessary to repeatedly determine V_o and V_m values. Normally the elution of small-scale analytical columns is performed by gravity flow; thus, flow-rate is determined by the hydrostatic head of the column. Brisk flow-rates of up to $20-60$ ml/h have been used.

(iv) Although V_o and V_m depend on the bed volume of matrix, these parameters as normally determined also include the volume inside the tubing connecting the column to the fraction collector, or detector if this is interposed between column and collector. Once V_o and V_m have been determined for a particular system, the physical configuration of column, buffer reservoirs, sample injector, connecting tube, detector and fraction collector should not be changed.

(v) The total penetrable volume of the column, V_o, is measured by elution of a molecule that is small enough to penetrate the gel matrix to the same degree as mobile interactant and has no specific affinity for the immobilised interactant. For analyses of small protein interactions on agarose-based affinity matrices such as those described below, bovine pancreatic ribonuclease or another similar common protein has been used to determine V_o.

(vi) The column void volume, V_m, is measured by elution of a large non-penetrating macromolecule; Blue Dextran has frequently been used for agarose-based matrices.

(vii) Non-biospecific interactions (such as ion-pairing between the mobile component and the matrix) leading to additional unwanted retardation should be avoided by appropriate selection of buffer and running conditions (5). The nature of non-specific interaction ultimately dictates the conditions needed to eliminate them.

(viii) The volume of sample applied to the column is usually kept small, typically below $100-400$ μl for columns with bed volumes of $1-10$ ml, respectively. Minimising starting sample volume minimises the width of the eluted peak and thus increases the accuracy of determining elution volume.

(ix) The amount of protein in the zone applied to the column should be minimal (see Section 3.1). Sensitive detection methods, such as radioisotope counting, radioimmunoassay and enzymatic assay, provide advantageous means for performing zonal elution analyses with small amounts of mobile interactant.

(x) Elution volumes can be determined visually by triangulation of the major peak. When unretarded contaminants elute at V_o and are detected by the assay method used, the resultant breakthrough peak can usually be ignored unless it interferes with analysis of the retarded peak of mobile interactant (elution volume close to V_o). Computer-assisted fit of elution data for determination of elution volumes offers a reliable alternative to triangulation.

3. ANALYSIS OF DATA CALCULATION OF EQUILIBRIUM CONSTANTS FROM ELUTION PROFILES

3.1 Monovalent Systems

The generalised scheme in *Figure 2* defines a set of competing monovalent binding reactions:

$$P + L \xrightleftharpoons{K_L} PL$$

Equation 1

$$P + \bar{L} \xrightleftharpoons{K_{\bar{L}}} P\bar{L}$$

where P is the mobile interactant, $\bar{L}$ is the matrix-immobilised interactant, L is a soluble component that competes with $\bar{L}$ for binding to P; PL and P$\bar{L}$ are non-covalent complexes formed between the mobile and immobilised species, respectively; and K_L and $K_{\bar{L}}$ are dissociation constants for PL and P$\bar{L}$, respectively. For the purposes of most of the present discussion, P is defined as a macromolecule such as a protein and L and $\bar{L}$ are small molecule ligands that bind to the same active site of P. However, the equations to be discussed are valid for other cases as well, e.g., when both P and $\bar{L}$ are macromolecules.

The equilibria in Equation 1 and considerations of interactive liquid chromatography (1,2,6) lead to the relationship:

$$\frac{1}{V - V_0} = \frac{K_{\bar{L}}}{(V_0 - V_m)[\bar{L}]} + \frac{K_{\bar{L}}[L]}{K_L(V_0 - V_m)[\bar{L}]}$$

Equation 2

where V_m and V_0 are as defined in Section 2. This expression relates the variation of V, the experimentally measured elution volume for mobile interactant, to the concentrations of immobilised and soluble competing ligands, $[\bar{L}]$ and $[L]$, respectively, and the dissociation constants $K_{\bar{L}}$ and K_L. Equation 2 allows $K_{\bar{L}}$ and K_L to be determined directly from chromatographic data for monovalent interacting systems. For a specific affinity matrix of fixed $[\bar{L}]$, a series of elutions is carried out in which $[L]$ is varied. The elution volumes of zones of mobile macromolecule are determined at the various values of $[L]$. From a plot of $1/(V - V_0)$ *versus* $[L]$, values for $K_{\bar{L}}$ and K_L can be calculated as defined by Equation 2. Thus, K_L can be derived from the ratio of slope/ordinate intercept and $K_{\bar{L}}$ from the ordinate intercept. Values of K_L and $K_{\bar{L}}$ also can be calculated non-graphically by linear least-squares regression analysis of the elution data. The values of $[\bar{L}]$ (see Section 4.2), V_0 and V_m are constants defined for a particular matrix and physical column arrangement.

In some instances, it may be desirable or necessary to perform zonal elutions in the absence of competing soluble ligand ($[L] = 0$). In this situation, Equation 2 simplifies to:

$$\frac{1}{V - V_0} = \frac{K_{\bar{L}}}{(V_0 - V_m)[\bar{L}]}$$

Equation 3

When $[L] = 0$, the value of $K_{\bar{L}}$ is directly related to $1/(V - V_0)$. Thus for the same matrix of fixed $[\bar{L}]$ that was used for competitive zonal elutions, a value for $K_{\bar{L}}$ can be calculated independently from a single elution profile.

An important condition of zonal analytical affinity chromatography is that the concentration of mobile interactant, [P], is indeterminate and continuously changing as the zone passes through the matrix bed. Thus, explicit terms involving [P] are not included in expressions relating elution volumes and dissociation constants (see equations above, for example). Neglecting [P] is valid only when this parameter is small with respect to the dissociation constant ($K_{\bar{L}}$) for binding of mobile and immobilised interactants. Carrying out zonal elution experiments at low [P] normally is easily achieved and, indeed, desirable when analysing bio-molecular species of finite availability. The use of low [P] in zonal elution analysis is analogous to the use of low [enzyme] in steady-state enzyme kinetic analysis (7). It has been shown experimentally (see for example ref. 5) that when the amount of enzyme applied to the affinity matrix is low relative to $K_{\bar{L}}$, the dependence of calculated dissociation constants on the amount of mobile interactant used is negligible. Alternatively, Nichol *et al.* (8) have defined equations for continuous elution/frontal analysis which accounts explicitly for the concentration of the mobile interactant applied to the affinity column (see also Chapter 6). From a practical point of view, the zonal elution approach is more convenient experimentally and requires far less mobile interactant. Thus, for many if not most interacting systems of biological interest, zonal elution analysis is likely to be the method of choice. Hence, the need for sensitive analytical methods to enable elution and detection of small amounts of mobile interactant.

3.2 Bivalent Systems

Zonal elution chromatography can also be applied to evaluate bivalent binding systems, including the type in Equation 4.

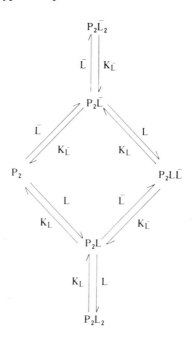

Equation 4

In a manner similar to that for monovalent binding systems (see 9,10) this scheme leads to the formulation:

$$\frac{1}{V - V_o} = \frac{1 + 2\frac{[L]}{K_L} + \left(\frac{[L]}{K_L}\right)^2}{(V_o - V_m)\left[2\frac{[\bar{L}]}{K_{\bar{L}}} + \left(\frac{[\bar{L}]}{K_{\bar{L}}}\right)^2 + 2\frac{[L]}{K_L}\frac{[\bar{L}]}{K_{\bar{L}}}\right]} \qquad \text{Equation 5}$$

This expression allows microscopic dissociation constants, K_L and $K_{\bar{L}}$, to be evaluated for a bivalent binding species, P_2, by measuring V at varying $[L]$. However, in contrast to monovalent binding systems, the variation of $1/(V - V_o)$ with $[L]$ for bivalent systems is non-linear. Thus, the values of K_L and $K_{\bar{L}}$ are derived from the competitive elution data by non-linear least-squares regression analysis. Despite this difference, the experimental protocol for collecting data for a bivalent binding system is the same as for a monovalent system.

When zonal elutions are carried out without soluble ligand present ($[L] = 0$), Equation 5 simplifies to:

$$V - V_o = (V_o - V_m)\left[2\frac{[\bar{L}]}{K_{\bar{L}}} + \left(\frac{[\bar{L}]}{K_{\bar{L}}}\right)^2\right] \qquad \text{Equation 6}$$

At $[L] = 0$, the value of $K_{\bar{L}}$ can be calculated directly from the elution volume of a single zonal elution.

4. AFFINITY MATRIX PREPARATION AND CHARACTERISATION

The development of affinity chromatography for preparative isolation of macromolecules has led to a considerable variety of techniques for immobilisation of ligands to solid supports. It is beyond the scope of this chapter to summarise the many methods developed. However, a few general principles for the design of affinity matrices to be used for analytical affinity chromatography should be noted.

4.1 Interactant Immobilisation

An important consideration for measuring equilibrium binding constants by zonal elution chromatography is to balance the capacity of the column with the affinity of the mobile component for the immobilised ligand in order to obtain well-formed elution profiles in realistic amounts of time. Obtaining well-formed elution profiles depends on parameters such as bead and pore size of the matrix, column diameter and length, and flow-rates (11,12). In addition, the retardation of mobile component depends directly on the amount of matrix-coupled ligand (1,3). When parameters such as flow-rate and column volume are kept constant, zones of mobile interactant applied to matrices containing lower amounts of immobilised ligand elute as sharper peaks at lower elution volumes. Increasing the amount of matrix-bound ligand leads to broader peaks and greater elution volumes. Trial and error is usually necessary to ascertain the appropriate immobilised ligand concentration for a particular series of experiments. The concentration of immobilised ligand coupled to a solid support can be controlled initially in the coupling procedure by reacting measured amounts of ligand with the activated solid support. A number of matrix preparations with different ligand densities can be tested subsequently to determine which concentration is best for a particular mobile interactant. Alternatively, the support can be reacted with a large amount of ligand and then diluted with underivatised matrix to achieve the desired immobilised ligand concentration. The latter procedure is less desirable when ligand density of the concentrated affinity matrix is large enough to affect mobile interactant retardation due to multivalent attachment or steric hinderance.

Coupling ligands to solid supports involves two steps, activation of the solid support and coupling between the ligand and the activated resin. In some instances, additional steps are incorporated to connect spacer arms of various lengths to the solid support. Ligands can then be coupled to the free end of the spacer arm. In general, overall strategies for coupling usually depend upon the immobilised ligand sought, the materials (support and ligand) available commercially, and the obvious need to maintain biospecificity and control capacity. Otherwise, the chemical strategies are essentially the same for preparing affinity matrices for analytical use as they are for preparative use.

4.2 Capacity Determination

When an affinity matrix is to be used analytically, the capacity of the derivatised material needs to be determined. The capacity of a substituted matrix can be defined functionally as the concentration of immobilised component available for interacting with the mobile component. For a 1:1 (monovalent) system, this can be determined as the amount of mobile component that binds at saturation per unit volume of the settled gel volume. When binding is not equimolar (e.g., for binding of mobile component multivalently to the immobilised component), the functional capacity of the matrix may be miscalculated (underestimated for multivalent mobile component). In addition, measurement of the capacity of a matrix by this method may not be practical if the affinity between the interacting components is too low to prevent premature elution of mobile component. Also, attempting to saturate a matrix may be prohibitive for a mobile binding component, especially a macromolecule or macromolecular assembly, which is in short supply. When it is not possible to determine the functional capacity of a matrix, the most appropriate value to report is the total amount of immobilised interactant per unit volume of matrix. Methods for determining matrix capacity are outlined below.

The functional capacity of a packed column containing the substituted solid support can be determined by saturating the packed column with the mobile binding component. An amount of mobile component in excess of the theoretical capacity of the derivatised support is applied to a measured amount of the matrix

(13). The matrix is washed with the loading buffer until the effluent is devoid of the mobile molecule. The bound substance is subsequently eluted and quantitated. When matrix saturation is not practical (for example, due to low affinity of the matrix for the mobile component), the binding capacity may be measured by a batch procedure. Here, small volumes of affinity matrix are mixed with varying concentrations of mobile component, the amount bound is determined at each concentration, and the resultant data are extrapolated to saturation. This allows estimation of the total amount of mobile component that can be bound at saturation of the matrix without actually saturating the matrix.

The total amount of immobilised component can be measured in two ways. The first is direct analysis of the solid support. For peptides and proteins, amino acid analysis after exhaustive acid hydrolysis (6.0 M HCl, *in vacuo*, 110°C, 24 h) has been used (e.g., ref. 4). For matrices containing immobilised nucleotides, phosphate analysis by the method of Fiske and Subbarow (14) following exhaustive hydrolysis of the derivatised matrix (15) has been used. An alternative is to determine the amount of ligand that is not covalently coupled to the solid support after the coupling reaction. This value is determined by subtracting the amount remaining in solution after reaction (including washes) from the starting amount of ligand.

4.3 Examples of Affinity Matrix Preparations

4.3.1 *pdTp-Aminophenyl-Sepharose*

Staphylococcal nuclease, derived from *Staphylococcus aureus*, hydrolyses RNA and DNA. Thymidine-3′,5′-bisphosphate (pdTp) is a powerful inhibitor of this enzyme and as such was chosen as the ligand for coupling to Sepharose for affinity chromatography of this enzyme (16). This matrix, pdTp-aminophenyl-Sepharose, is prepared by a one-step coupling reaction between cyanogen bromide-activated Sepharose and 3′-(4-aminophenylphosphoryl)-deoxy-thymidine-5′-phosphate (pdTpAP).

(i)   Wash cyanogen bromide (CNBr) activated Sepharose, which can be prepared (Section 1 of Chapter 2 and ref. 17) or purchased, with 0.1 M bicarbonate buffer, pH 9.0, in an equal volume of the same buffer.

(ii)   Add the ligand, pdTpAP, in a solution representing about 5 – 15% of the final volume.

(iii)   Stir the suspension gently (so as not to damage the Sepharose beads) at 4°C for 24 h, and then wash extensively with water and buffer.

The amount of the ligand coupled to cyanogen bromide-activated Sepharose under the conditions described above has been found on one occasion to be 1.5 and 0.3 μmol of immobilised ligand/ml Sepharose when, respectively 2.5 and 0.5 μmol of soluble ligand were added initially per ml of Sepharose (16). An empirical quantitation of the functional capacity has been made by saturation of a measured amount of the ligand with pure staphylococcal nuclease. The bound protein was eluted and quantitated by an activity assay. For the matrices found to contain 1.5 and 0.3 μmol of ligand/ml Sepharose, the amounts of nuclease reported to bind were 8 and 1.2 mg/ml Sepharose, indicating [$\bar{\text{L}}$] values (by capacity) of 0.5 and 0.08 mM, respectively. Based on the total phosphate content of the matrices, the theoretical maximum binding would have been 28 and 6 mg/ml Sepharose, respectively, assuming 1:1 binding of the enzyme to total immobilised ligand (16). For the analytical elutions on pdTpAP-Sepharose shown in Section 6.1 [$\overline{L}$] = 0.05 mM, as measured by functional capacity, *versus* 0.21 mM by total phosphate content.

4.3.2 *Phosphorylcholine-Sepharose*

A bivalent IgA antibody (TEPC 15), isolated from BALB/c mouse gamma-A myelomas, is known to bind specifically to phosphorylcholine. Phosphorylcholine linked to Sepharose has been used successfully as an affinity matrix both for preparative (18) and quantitative affinity chromatography (9). The affinity matrix has been prepared by a two-step procedure (18). In the first step, glycyltyrosine is coupled to CNBr-activated Sepharose through the α-amino group of glycine; the second step is a substitution reaction in which the diazo group of *p*-diazoniumphenylphosphorylcholine attacks the ring *ortho-* to the hydroxyl group of the phenol ring of the matrix-attached glycyltyrosine. For analytical purposes, this substituted matrix may be prepared by first reacting a fixed amount of CNBr-activated Sepharose (10 ml wet weight volume) with 35 μmol of glycyltyrosine (9). The preparation of CNBr-activated Sepharose and the subsequent reaction with glycyltyrosine is according to published procedures (17). After the first coupling reaction, wash the substituted resin with borate-buffered saline, pH 8.0, then suspend in the same buffer and react with 5 μmol of *p*-diazoniumphenylphosphorylcholine. Stir this mixture at room temperature overnight. After the reaction, wash the substituted resin with water and store in 1.0 M acetic acid at 4°C.

The total amount of phosphorylcholine coupled to the matrix is determined as total phosphate. For the matrix prepared with 35 μmol of Gly-Tyr/10 ml activated Sepharose, phosphate content was determined to be 1.0×10^{-3} M. The functional capacity was measured by saturating 0.1 ml of the substituted resin with 2 mg of IgA monomer containing radioactively-labeled antibody as tracer. The column was washed with phosphate-buffered saline (PBS) containing 1 mg/ml bovine serum albumin until unbound protein was eluted completely. Antibody specifically bound to the column was eluted with buffer containing 10^{-3} M phosphorylcholine. The capacity of the phosphorylcholine-Sepharose for IgA antibody, taken as the amount of antibody eluted at this step, was 5×10^{-5} M (9).

A more diluted gel may be made for analytical use, by reacting 3.5 μmol Gly-Tyr with 10 ml activated Sepharose. Here the functional capacity was 9×10^{-6} M (9). Both this diluted and the above more concentrated matrices were used for analytical elutions shown in Section 6.2.

4.3.3 *Methionyl-tyrosyl-phenylalanyl-aminoalkyl-agarose*

Neurophysins, a family of 10 000-dalton, acidic proteins isolated originally from the posterior pituitary, bind the neuropeptide hormones, oxytocin and vasopressin non-covalently. Extensive studies have revealed that the N-terminal three amino acid residues of the hormones (half-cystine is the N-terminal residue, followed by tyrosine and either phenylalanine in vasopressin or isoleucine in oxytocin) are the most important for the protein-peptide binding interaction (19). The tripeptide amide, Met-Tyr-Phe amide, is an effective ligand for the hormone binding site of neurophysins.

Coupling of the tripeptide (free acid) to an agarose solid support through aminoalkyl spacer arms has been shown to provide an excellent and convenient affinity matrix for both preparative and quantitative affinity chromatography of neurophysins (4,20).

Met-Tyr-Phe-Aminohexyl-agarose may be prepared in a one-step coupling reaction between *p*-nitrophenylsulphenyl-Met-Tyr-Phe (NPS-Met-Tyr-Phe) and ω-aminohexyl-agarose (20).

(i)   Add NPS-Met-Tyr-Phe (20 mg in 5.5 ml dioxane) to 5 ml (settled gel volume in water) of ω-aminohexyl-agarose.

(ii)   Initiate the coupling reaction by adding 33 μmol of both dicyclohexyl-carbodiimide and N-hydroxysuccinimide (1 μmol/μl in DMF).

(iii) After 20 h at room temperature, filter the substituted gel and wash sequentially with dioxane-water (1:1 v/v), methanol-water (1:1 v/v), water, 1.0 M NaCl, and water.

(iv)   Suspend the resulting yellow resin in 0.2 M ammonium acetate, pH 5.2, containing 0.2 M sodium thiosulphate and mix for 1 h at room temperature to remove the NPS protecting group from the amino-terminal end of the peptide.

(v)   Subsequently filter the gel and wash with water, methanol, water, DMF-water (1:1 w/v), and water.

Amino acid analysis of the matrix prepared in the proportions described above, reveals that 0.85 μmol of Met-Tyr-Phe tripeptide is incorporated per ml of settled gel.

For the affinity matrices used in the analytical experiments in Section 6.3, the concentration of immobilised ligand for Met-Tyr-Phe-aminohexyl and aminobutyl-agarose was determined to be 0.59 and 0.30 mM, respectively, by amino acid analysis (4).

4.3.4 *BNP-II-Sepharose*

In addition to the ligand binding activity of neurophysins (as described above), this family of proteins also self-associate to form dimers from monomer subunits. This interaction has been studied quantitatively by affinity chromatography using a matrix of bovine neurophysin II (BNP-II) coupled to Sepharose (4,21,22). The BNP-II-Sepharose matrix may be made by a one-step coupling reaction between CNBr-activated Sepharose and BNP-II in 0.1 M sodium bicarbonate, pH 8.5, containing 0.5 M sodium chloride, overnight at 4°C, with gentle stirring. Wash the reacted gel with coupling buffer and mix with 1.0 M ethanolamine in coupling buffer for 3 h at room temperature, followed by washings with coupling buffer and 0.1 M acetate - 1 M sodium chloride, pH 3.5.

The substituted matrix used in the analytical experiment in Section 6.4 was obtained (21) from reaction of 1 g activated Sepharose with 5 mg (by weight) of BNP-II and contained 1.4 mg BNP-II/ml packed gel as determined by amino acid analysis.

5. PREPARATION OF SAMPLES FOR QUANTITATIVE AFFINITY CHROMATOGRAPHY AND ELUTION MONITORING

5.1 Sample Preparation and Radioisotopic Labelling

5.1.1 *General Comments*

Samples to be used for quantitative affinity chromatography should be maximally homogeneous, and are therefore often purified by preparative affinity chromatography before being used for analytical scale chromatography. Beyond this, the preparation of a sample is case-specific. In cases for which a sensitive assay (e.g., enzymatic activity or radioimmunoassay) cannot be used for detection of mobile interactant, radioisotope labelling is advantageous. The major criteria for deciding which method to use for radioisotope labelling of mobile interactant are: (i) achieving sufficient label incorporation to allow elution and subsequent detection of small amounts of the component; and (ii) maintaining normal, biospecific binding by the labelled species. Incorporation of ^{125}I at amino groups by N-succimidyl-3-(4-hydroxy-5-[^{125}I]iodophenyl) propionate (23), tritium by reductive methylation (24), and [^{14}C]carbamylation (25) all have been used successfully as general methods for labelling proteins for quantitative affinity chromatography.

5.1.2 *Reductive Alkylation*

Tritiated IgA antibody may be prepared by reductive methylation of monomeric IgA (9). To accomplish this, dissolve monomeric IgA in 0.2 M borate, pH 8.3 (2 mg reduced and alkylated IgA in 2 ml buffer). To this add 40 mCi of sodium [^{3}H]borohydride (10 Ci/mmol, obtained from New England Nuclear Inc.). Add 20 μl formaldehyde as a 0.4% solution to initiate the methylation reaction (10 min total reaction time, on ice). Remove unreacted reagents at the end of the reaction by gel filtration. Finally, dialyse the tritium-labeled protein against PBS, pH 7.4. The resulting specific activity of the tritiated protein prepared in the proportions described above (used for experiments described in Section 6.2) is approximately 0.5 μCi/μg.

5.1.3 *Carbamylation*

Carbon-14 labelled bovine neurophysin I ([^{14}C]BNP-I) has been prepared (22) by carbamylation of native BNP-I with ^{14}C-labelled potassium cyanate (KCNO, specific activity 6.2 mCi/mmol). To accomplish this, dissolve BNP-I (12 mg) in 0.5 M borate, pH 8.5 (0.75 ml), and react with 12 mg ^{14}C-labelled KCNO for 15 h at room temperature. When the reaction time is complete, reduce the pH of the solution to 7.4 with glacial acetic acid and fractionate the mixture by gel filtration chromatography (Sephadex G-50, 0.9 x 58 cm, equilibrated in PBS, pH 7.4). The specific activity of BNP-I radiolabelled in the proportions described above is about 1 nCi/μg protein. Such labelled neurophysin is shown to be active by its ability to bind to lysine vasopressin-Sepharose (22) and has been used in analytical affinity chromatography (20).

5.1.4 *Iodination*

^{125}I-Labelled BNP-II may be prepared (4) by reaction with N-succinimidyl 3-(4-hydroxy-5-[^{125}I]iodophenyl)propionate according to the procedure of Bolton and Hunter (23). Typically, add a 20 μl aliquot of BNP-I or II (5 mg/ml in ammonium acetate, pH 5.7) and 40 μl of 1.0 M borate, pH 8.5 to the dried Bolton-Hunter reagent (1 mCi, 2000 Ci/mmol, from New England Nuclear). After 15 – 30 min on ice with frequent agitation, react excess reagent for 5 – 10 min on ice, with 300 – 500 μl of 0.2 M glycine in 0.5 M borate, pH 8.5. After the reaction, separate [^{125}I]BNP-I or II from reagents by Sephadex G-25 (PD-10, Pharmacia) chromatography in 0.2 M acetic acid and then, after removal of acid, repurify by preparative affinity chromatography on Met-Tyr-Phe-aminohexyl-agarose (20). Radiolabelled material specifically bound to the matrix in 0.4 M ammonium acetate, pH 5.7, and eluted in 0.2 M acetic acid may be lyophilised in the presence of added crystalline bovine serum albumin. Dissolve the dried, labelled protein in 10 mM phosphate, pH 7.4, to make a stock solution. A preparation of [^{125}I]BNP-II as used for experiments in Section 6.3 has a specific activity of 2.5 μCi/nmol protein (4), while that for elutions shown in Section 6.4 is 0.3 μCi/nmol (21).

5.2 **Elution Monitoring**

5.2.1 *Enzyme Assay*

Studies of enzyme-immobilised inhibitor binding interactions can utilise the enzymatic activity of the enzyme to monitor elution of the zone from the matrix. This approach offers the advantage that the sample does not have to be covalently derivatised. On the other hand, when competitive elutions are performed, the fractions containing the eluted enzyme may need to be diluted or dialysed first to reduce interference of enzymatic activity by soluble competing ligand.

5.2.2 *Radioactivity Measurement*

Radiolabelled samples can be detected at high sensitivity. Thus, when other more intrinsic molecular activities are not sufficiently sensitive, radioisotope incorporation makes analysis possible for the small amounts of mobile component desired for zonal elution affinity chromatography (see Sections 2 or 3.1). The method for detection of radiolabelled samples is straightforward. Carbon-14 and tritium-labelled samples are detected by scintillation counting of samples obtained by mixing a fixed aliquot of scintillation cocktail with a consistent amount (to maintain equal quenching by buffer components) of each fraction collected during a column elution. The advent of in-line scintillation counting offers the possibility to eliminate sample aliquoting, although, at least with current instrumentation, sensitivity of detection is reduced.

Iodine-125-labelled samples offer the most convenient and sensitive radiolabel detection for elution monitoring. Eluted fractions can be inserted directly into an automatic gamma counter. Such ease of sample handling and data collection permits the investigator to generate many column elution profiles in a short time. Samples that are labelled with iodine-125 to moderate specific activities $(5-500 \text{ nCi}/\mu\text{g})$ can be detected easily and reliably when applied in submicrogram amounts.

5.2.3 *Other Techniques*

When enzyme assays or direct radioactivity measurements are not appropriate for elution monitoring, other types of binding assays may be considered.

Perhaps most notably, radioimmunoassay provides a sensitive general approach for quantitation of the mobile component. Also, estimating the amount of eluted mobile component by a quantitative ligand-binding assay may be realistic, for example, if a soluble radiolabelled ligand analogous in structure to the immobilised ligand is available.

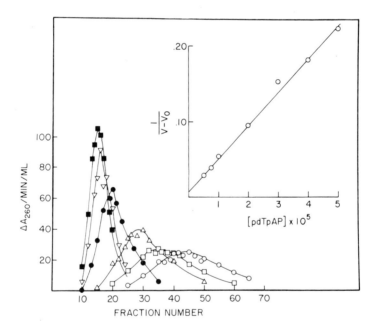

Figure 4. Competitive zonal elution affinity chromatography for a monovalent binding system. Zones containing equal amounts of S. nuclease were applied to a pdTpAP-Sepharose column (0.9 x 15 cm, $[\bar{L}] = 0.05$ mM) equilibrated with 0.1 M ammonium acetate, pH 5.7, containing soluble competitive ligand (pdTpAP) at concentrations of 0.5×10^{-5} M ($\bigcirc$), 0.785×10^{-5} M ($\square$), 1.0×10^{-5} M ($\triangle$), 2×10^{-5} M ($\bullet$), 3×10^{-5} M (∇), and 4.0×10^{-5} M ($\blacksquare$) and eluted with these same buffers. Elutions were carried out at room temperature. **Inset:** data from the main figure (and additional data at 5×10^{-5} M [pdTpAP] omitted from the main figure to avoid clutter) are plotted as $1/(V - V_o)$ *versus* [pdTpAP]. Dissociation constants, K_L and $K_{\bar{L}}$, were calculated from the linear plot using Equation 2 (see Section 3.1) and are listed in *Table 1*. Taken from Ref. 3.

6. EXAMPLES OF ZONAL ELUTION ANALYTICAL AFFINITY CHROMA-TOGRAPHY RESULTS

Several protein-ligand and protein-protein interacting systems have been studied by zonal elution quantitative affinity chromatography. A few representative examples are given below, showing the types of chromatographic results obtained for monovalent (e.g., staphylococcal nuclease/nucleotide), bivalent (e.g., TEPC 15 IgA bivalent monomer/phosphorylcholine), and cooperative (e.g., BPN-II/ peptide hormone) interacting complexes.

6.1 Staphylococcal Nuclease Binding to Immobilised Nucleotides

The elutions in *Figure 4* were obtained when zones of staphylococcal nuclease were eluted on pdTpAp-Sepharose at different concentrations of the soluble competitor (3). The linearity of the $1/(V - V_o)$ *versus* [L] plot reflects the 1:1 nature of the enzyme/nucleotide interaction and allows calculation of the dissociation constants K_L and $K_{\bar{L}}$ using the monovalent model, Equation 2. The calculated constants are given in *Table 1* which also contains results for two other competitors, connoting the potential of the method to determine straightforwardly equilibrium binding constants for a series of competitors and therein to characterise specificity properties of the interaction.

6.2 Bivalent Immunoglobulin A (TEPC 15 IgA monomer)/Phosphorylcholine

When TEPC 15 IgA, which has been reduced and alkylated selectively to yield bivalent monomers, was eluted on immobilised phosphorylcholine at 5×10^{-5} M immobilised ligand concentration (determined by functional capacity), competitive elutions with soluble phosphorylcholine could be achieved but the response of $1/(V - V_o)$ to the soluble competitor concentration was non-linear (9). This behaviour, shown in *Figure 5*, is consistent with the potential for bivalent binding of IgA monomer to the affinity matrix. As a result, the interaction parameters K_L and $K_{\bar{L}}$ were evaluated using the bivalent model Equation 5. These values are given in *Table 1*. When bivalent IgA is eluted on phosphoryl-choline-Sepharose at 9×10^{-6} M immobilised ligand concentration, variation of elution volume with soluble phosphorylcholine conforms, as shown in *Figure 6*, to the monovalent model. Here, the reduced density of matrix binding sites eliminates any significant chance for simultaneous binding of protein to two matrix sites. Thus, the monovalent as well as bivalent models can be used to calculate K_L and $K_{\bar{L}}$ values as given in *Table 1*.

While the IgA monomer elution on the higher concentration affinity matrix (5×10^{-5} M) leads to curvilinear $1/(V - V_o)$ *versus* [L] profiles, elution of monovalent IgA-derived Fab fragments on the same matrix shows linear behaviour, as seen in *Figure 7*. This verifies the interpretation that the curvilinearity with IgA monomers indeed results from incipient bivalent binding to the affinity matrix when the binding sites are close together (high density substitution).

Table 1. Representative Affinity Matrix Characteristics, Elution Analysis and Calculation Methods, and Quantitative Results Obtained in Zonal Elution Quantitative Affinity Chromatography.

Mobile interactant	Amount applied in zone	Affinity matrix	Method of $[\bar{L}]$ determination	Effector[a]	$[\bar{L}]$ (M)	Eluted mobile interactant detection method	Equation for calculating K_L and $\bar{K}_L$	K_L (M)	$\bar{K}_L$ (M)	Reference
Staphylococcal nuclease	200 μg	pdTpAP-Sepharose	functional capacity	pdTpAP	5×10^{-5}	enzymatic activity	2	2.3×10^{-6}	1.1×10^{-6}	(3)
				pdTp	9×10^{-6}	enzymatic activity	2	2.5×10^{-6}	1.0×10^{-6}	(3)
				dTpNP		enzymatic activity	2	4.3×10^{-3}	1.5×10^{-6}	(3)
[14C]IgA bivalent monomer[b]	25 μg	phosphorylcholine-Sepharose	functional capacity	phosphoryl-choline	5×10^{-5}	scintillation counting	5	1.7×10^{-6}	2.7×10^{-6}	(9)
					9×10^{-6}		5	1.6×10^{-6}	4.8×10^{-6}	(9)
							2	1.2×10^{-6}	1.2×10^{-6}	(9)
[3H]IgA monovalent Fab[c]	0.04 μg	phosphorylcholine-Sepharose	functional capacity	phosphoryl-choline	5×10^{-5}	scintillation counting	2	3.3×10^{-6}	4.2×10^{-6}	(9)
					9×10^{-6}		2	1.5×10^{-6}	3.9×10^{-6}	(9)
[125I]BNPII[d]	<1 μg	Met-Tyr-Phe-amino-hexyl-agarose	amino acid analysis	OXT	5.9×10^{-4}	γ-counting	2	6.6×10^{-5}	4.7×10^{-5} (at $[L] \neq 0$)	(4)
				—		γ-counting	3	—	4.8×10^{-5} (at $[L] = 0$)	(4)
	101 μg	Met-Tyr-Phe-amino-hexyl-agarose	amino acid analysis	OXT	5.9×10^{-4}	γ-counting	2	1.8×10^{-6}	7.0×10^{-7} (at $[L] \neq 0$)	(4)
				—		γ-counting	3	—	6.1×10^{-6} (at $[L] = 0$)	(4)
	<1 μg	BNPII-Sepharose	amino acid analysis	None	1×10^{-4}	γ-counting	3	—	1.4×10^{-5}	(21)
				LVP (1×10^{-4} M)		γ-counting	3	—	4.4×10^{-7}	(21)

[a] All effectors listed are competitive ligands, except for LVP which is a positive effector; OXT = oxytocin; LVP = lysine, vasopressin; pdTp = 3',5'-deoxythymidine diphosphate; dTpNP = deoxythymidine-3'-p-nitrophenyl phosphate; pdTpAP = deoxythymidine-5'-phosphate-3'-p-aminophenyl phosphate.

[b] TEPC 15, IgA which was reduced gently and alkylated with [14C]iodoacetamide.

[c] Fab fragments derived from TEPC 15 IgA monomers by limited papain digestion of protein tritiated by reductive methylation.

[d] BNPII reacted with [125I]Bolton-Hunter reagent.

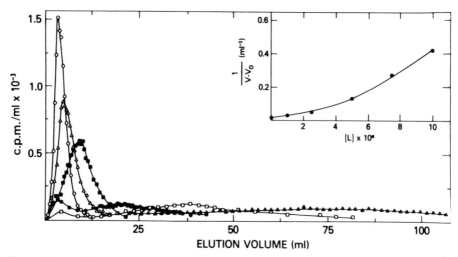

Figure 5. Competitive zonal elution affinity chromatography for a bivalent binding system. Zones (100 μl) of [^{14}C]IgA monomer (bivalent binding) were applied to high density phosphorylcholine-Sepharose (7 x 25 mm, [$\bar{L}$] = 5 x 10^{-5} M) equilibrated in PBS (with 1 mg/ml bovine serum albumin) and containing soluble phosphorylcholine at the following concentrations (M); 0 (▲), 1 x 10^{-6} (□), 2.5 x 10^{-6} (●), 5.0 x 10^{-6} (■), 7.5 x 10^{-6} (△), 1.0 x 10^{-5} (○). Elutions were carried out at room temperature with buffer containing the indicated amount of soluble competitive phosphorycholine. Inset: elution data are plotted as $1/(V - V_o)$ *versus* [phosphorylcholine]. Dissociation constants, K_L and $K_{\bar{L}}$, were calculated from the non-linear plot using Equation 5 and are shown in *Table 1*. (Taken from ref. 9).

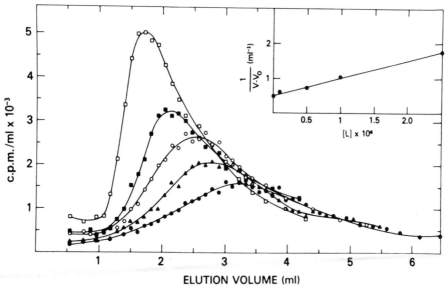

Figure 6. Competitive zonal elutions of bivalent IgA monomers on low density phosphorylcholine-Sepharose. Zones of [^{14}C]IgA monomer were applied to low density phosphorylcholine-Sepharose (7 x 25 mm, [$\bar{L}$] = 9 x 10^{-6} M) equilibrated and eluted as described for the high density phosphorylcholine-Sepharose column in *Figure 5* except that the following concentrations of soluble phosphorylcholine were used (M): 0 (●), 1 x 10^{-7} (▲), 5 x 10^{-7} (○), 1 x 10^{-6} (■), and 2.5 x 10^{-6} (□). Inset: variation of elution volumes (V) with concentration of free phosphorylcholine ([L]) plotted as $1/(V - V_o)$ *versus* [L]. Taken from ref. 9.

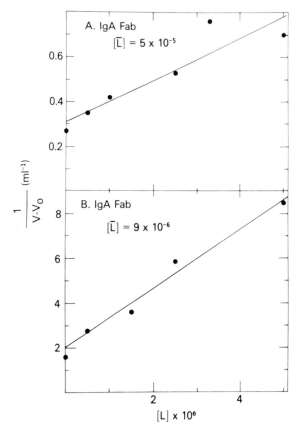

Figure 7. Elution behaviour of [³H]IgA Fab fragments on high density **(A)** and low density **(B)** phosphorylcholine-Sepharose columns, plotted as the variation of $1/(V - V_o)$ with soluble competing phosphorylcholine concentration. Chromatographic conditions for elutions of zones of [³H]IgA Fab fragments are as described in *Figure 5* for IgA monomer. Dissociation constants, K_L and $K_{\bar{L}}$, were calculated from the linear plots using Equation 2 and are shown in *Table 1*. Taken from ref. 9.

6.3 Bovine Neurophysin II/Peptides

Zonal elution of BNP-II on Met-Tyr-Phe-aminoalkyl-agaroses shows behaviour which fits a cooperative binding scheme distinct from the strictly monovalent or bivalent models. Elution volumes do decrease generally as the concentration of competitive peptide (e.g., oxytocin or vasopressin) is increased in the elution buffer, as shown for example in *Figure 3*. However, when sets of competitive elution data were plotted as $1/(V - V_o)$ *versus* [L], curvilinearity at low [L] was observed in most cases (*Figure 8*). The reason for this behaviour rests with the fact that neurophysin exists as a mixture of low affinity monomers and high affinity dimers, with the degree of dimerisation increased by peptide ligand binding. Thus, while elution of neurophysin at zero soluble ligand reflects the affinity matrix binding of a mixture containing a significant amount of monomers, the elution in the presence of low concentrations of soluble peptide is more retarded

186

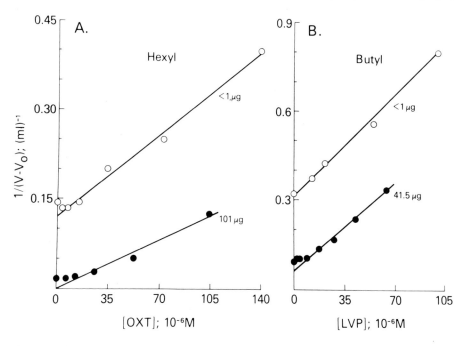

Figure 8. Linearised plots of competitive zonal elution data for [125I]-labelled bovine neurophysin II ([125I]BNP-II) on Met-Tyr-Phe-aminohexyl- **(A)** and -aminobutyl- **(B)** agarose (columns were 2 ml bed volumes packed in 7 mm internal diameter columns). Elution volumes (V) for the elution series shown in *Figure 3* and three other similar series (not shown) are plotted as $1/(V - V_o)$ *versus* the concentrations (as indicated in the figure) of competitive ligand ([L]) which in this case is either of the neuropeptide hormones, oxytocin (OXT) or lysine-vasopressin (LVP). Open circles are for elutions at low total protein in zones applied to the column (< 1 μg of [125I]BNP-II only); closed circles are for high total amounts of protein applied to the column (1 μg of [125I]BNP-II plus either 101 μg or 41.5 μg of unlabelled BNP-II). Straight lines were drawn according to linear least-squares regression analysis of data points except for points at [L] = 0 and for values of [L] near 0 which deviate from linearity. Dissociation constants, K_L and $K_{\bar{L}}$, calculated from the linearised plots using Equations 2 and 3 are shown in *Table 1*. Taken from ref. 4.

than expected; this reflects elution of mixtures progressively more enriched in high affinity dimers. Elution volumes decrease more gradually, with increasing but low soluble peptide concentrations, than expected from the degree of competition of the soluble peptides with affinity matrix for binding to the mobile neurophysin (4).

The higher peptide affinity of dimers than monomers is also expressed as a strong dependence of neurophysin elution volume on the amount (i.e., concentration) of neurophysin in the initial zone. Thus, as shown in *Figure 9*, increasing zonal protein concentrations leads to increasing retardation.

6.4 Neurophysin/Neurophysin

Neurophysin dimerisation has been measured by elution of neurophysin on immobilised neurophysin. As shown in the inset to *Figure 10*, zonal elution of [125I]BNP-II in buffer (containing no soluble peptide ligand) shows a retardation

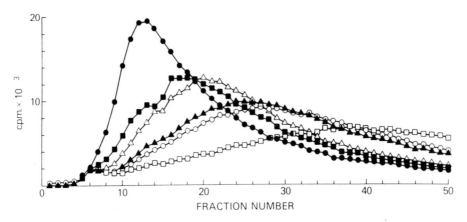

Figure 9. Effect of concentration of applied bovine neurophysin II (BNP-II) on zonal elution behaviour on Met-Tyr-Phe-aminobutyl-agarose. Zones (100 μl) containing < 1 μg of [^{125}I]BNP-II and unlabelled BNP-II (varying amounts as specified) were eluted in 0.4 M ammonium acetate, pH 5.7, in the absence of soluble competitive ligand. Each continuous profile represents a separate elution with the following amounts (in micrograms) of added unlabelled BNP-II per zone: 0 (●), 6.35 (■), 10.3 (△); 20.7 (▲), 41.3 (○), 82.6 (□). Elutions were carried out at room temperature. Taken from ref. 4.

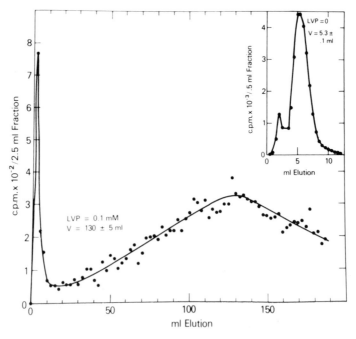

Figure 10. Elution behaviour of [^{125}I]BNP-II on [BNP-II]-Sepharose in the presence and absence of soluble peptide ligand, lysine-vasopressin (LVP). Zones (300 μl) containing ~1 μg of [^{125}I]BNP-II were applied to [BNP-II]-Sepharose (1.5 ml bed volume, [$\bar{L}$] = 1.4 x 10^{-4} M) equilibrated with 0.4 M ammonium acetate, pH 5.7, containing 0.1 mM LVP (main diagram) or in the absence of added LVP (inset). The zonal elution volumes were used to calculate $K_{\bar{L}}$ (dissociation constants for protein-protein interaction) using Equation 3. These values are shown in *Table 1*. Taken from ref. 21.

displaced sufficiently from V_0 to allow calculation of a dimerisation constant as K_L^- (see *Table 1*). By including soluble peptide ligand in the elution buffer, dimerisation dissociation constants can also be determined for partially or fully liganded neurophysin. The main elution profile of *Figure 10* shows the case of 'close-to-saturating' lysine-vasopressin. For this liganded case, K_L^- is much smaller (higher affinity) than that in buffer alone (see *Table 1*). Thus, here as with the peptidyl affinity matrix, zonal elution of neurophysin can be used to measure the degree of modulation of binding affinities in the cooperative complexes between neurophysins and peptide hormones.

7. CONCLUDING COMMENT

Zonal elution analytical affinity chromatography has been evaluated with a variety of macromolecular systems. The method has been found to be an advantageous, technically straightforward way to characterise interactions of such molecules. Its growing use in several laboratories (26) reflects the potential benefits of the technique in studying biomolecular interactions and therein molecular bases of function in biological systems.

8. REFERENCES

1. Dunn,B.M. and Chaiken,I.M. (1974) *Proc. Natl. Acad. Sci. USA,* **71**, 2382.
2. Chaiken,I.M. (1979) *Anal. Biochem.,* **97**, 1.
3. Dunn,B.M. and Chaiken,I.M. (1975) *Biochemistry (Wash.),* **14**, 2343.
4. Angal,S. and Chaiken,I.M. (1982) *Biochemistry (Wash.),* **21**, 1574.
5. Chaiken,I.M. and Taylor,H.C. (1976) *J. Biol. Chem.,* **251**, 2044.
6. Hethcote,H.W. and De Lisi,C. (1982) *J. Chromatogr.,* **248**, 183.
7. Bender,M.L. and Brubacher,L.J. (1973), *Catalysis and Enzyme Action*, published by McGraw-Hill Book Co., New York, p. 25.
8. Nichol,L.W., Ogston,A.G., Winzor,D.J. and Sawyer,W.H. (1974) *Biochem. J.,* **143**, 435
9. Eilat,D. and Chaiken,I.M. (1979) *Biochemistry (Wash.),* **18**, 790.
10. DeLisi,C., and Hethcote,H.W. (1982) in *Affinity Chromatography and Related Techniques*, Gribnau,T., Visser,J. and Nivard,R. (eds.), Elsevier Scientific Publishing Co., Amsterdam, p. 63.
11. Hethcote,H.W. and DeLisi,C. (1982) *J. Chromatog.,* **240**, 269.
12. DeLisi,C., Hethcote,H.W. and Brettler,J.W. (1982) *J. Chromatogr.,* **240**, 283.
13. Wilchek,M. and Gorecki,M. (1969) *Eur. J. Biochem.,* **11**, 491.
14. Fiske,C.H. and Subbarow,Y. (1925) *J. Biol. Chem.,* **66**, 375.
15. Bartlett,G.R. (1959) *J. Biol. Chem.,* **234**, 466.
16. Cuatrecasas,P., Wilchek,M. and Anfinsen,C.G. (1968) *Proc. Natl. Acad. Sci. USA,* **61**, 636.
17. Cuatrecasas,P. (1970) *J. Biol. Chem.,* **245**, 3059.
18. Chesebro,B. and Metzger,H. (1972) *Biochemistry (Wash.),* **11**, 766.
19. Breslow,E., Weis,J. and Menendez-Botet,C.J. (1973) *Biochemistry (Wash.),* **12**, 4644.
20. Chaiken,I.M. (1979) *Anal. Biochem.,* **97**, 302.
21. Chaiken,I.M., Tamaoki,H., Brownstein,M.J. and Gainer,H. (1983) *FEBS Lett.,* **164**, 361.
22. Fischer,E.H., Curd,J.G. and Chaiken,I.M. (1977) *Immunochemistry,* **14**, 529.
23. Bolton,A.E. and Hunter,W.M. (1973) *Biochem. J.,* **133**, 529.
24. Rice,R.H. and Means,G.E. (1971) *J. Biol. Chem.,* **246**, 831.
25. Stark,G.R. (1967) *Methods Enzymol.,* **11**, 125.
26. Chaiken,I.M., Wilchek,M. and Parikh,I., eds. (1983) *Affinity Chromatography and Biological Recognition*, Academic Press, New York.

CHAPTER 8

Cell Separation by Affinity Chromatography Ligand Immobilisation to Solid Supports Through Cleavable Mercury-Sulphur Bonds

J.C. BONNAFOUS, J. DORNAND, J. FAVERO and J.-C. MANI

1. INTRODUCTION

The understanding of many biological phenomena requires isolation of cell sub-populations which are in various stages of differentiation and forms of specialisation. Cell separations are especially important in the study of the immune system which consists of a complex array of cells possessing specific phenotypes and functions. The principal methods are based on the presence of specific components on the surface of the cells to be purified: e.g., lectin receptors, antigens, immunoglobulins. Their development has been greatly facilitated by the increasing availability of monoclonal antibodies against cell surface determinants.

Cell isolation with fluorescence activated cell sorters (FACS) requires expensive equipment and can hardly be applied to routine experimentation, especially if great numbers of cells are needed. Consequently there is a need for more widely applicable and reproducible methods for cell separation. Affinity chromatography constitutes in principle an ideal method for purification of cells: it consists of immobilisation on solid supports of ligands which selectively recognise membrane components.

2. PRESENT STATE OF CELL AFFINITY CHROMATOGRAPHY

Affinity techniques have sometimes been used as a means of negative selection, i.e., elimination of a specific cell population. However, the most interesting methodology, related to classical affinity chromatography, involves positive immunoselective procedures. Numerous applications have been attempted during the last 15 years; these have been reviewed elsewhere (1 – 4). This chapter only outlines the principles of the most recent methods (*Table 1*) among which three show the widest potential application.

(i) *Rosette techniques* (4 – 8): fixation of anti-mouse immunoglobulins on ox red blood cells (by the chromic chloride method) and rosetting with the cells to be purified which have previously been coated with specific monoclonal antibodies (produced by mouse hybridoma); recovery of bound cells is achieved by selective lysis of the erythrocytes.

(ii) *Panning techniques* (9 – 15): adsorption of antibodies on polystyrene dishes (in a chemically undefined manner); most often double antibody techniques are used, i.e., plastic dishes coated with anti-mouse immunoglobulin antibodies selectively bind cells previously treated with specific monoclonal antibodies.

Table 1. Some Recent Examples of Cell Separation by Affinity Chromatography.

Matrix	Immobilised ligand	Cells specifically adsorbed on the affinity support	Means of recovery of bound cells	Reference
Ox red blood cells	Rabbit anti-mouse IgG antibodies (coupled to ox red blood cells with chromic chloride)	Cells bearing HLA-DR antigens, coated with anti-HLA-DR monoclonal antibodies	Ammonium chloride selective lysis of erythrocytes	5
Ox red blood cells	Rabbit IgG anti-mouse IgG	$T_8{}^+$ (suppressor) and Leu-3a$^+$ (helper) cells previously coated with OKT_8 and anti-Leu-3a monoclonal antibodies	Lysis of ox red blood cells	6
Ox red blood cells	Anti-mouse IgG antibodies	Human lymphocyte subsets coated with specific mouse monoclonal antibodies	Lysis of ox red blood cells	7
Ox red blood cells	Rabbit anti-mouse Ig antibodies coupled to ox red blood cells through SpA and rabbit anti-ox antibodies	Human lymphocyte subsets coated with specific mouse monoclonal antibodies	Only detection experiments described	8
Plastic plates	− Anti-mouse Ig antibodies − F(ab′)$_2$ fragments of anti-human Ig antibodies	− Isolation of human T lymphocyte subsets coated with specific monoclonal antibodies (OKT_4, OKT_8) − Fractionation of human B and T lymphocytes	Gentle hydrostatic pressure (pipetting)	9
Plastic dishes	Goat anti-mouse immunoglobulin antibodies	Human lymphocyte subsets ($T_4{}^+$ and $T_8{}^+$) coated with specific mouse monoclonal antibodies	Mechanical	10
Plastic dishes	Goat anti-mouse immunoglobulins (Fc portion) antibodies	Human lymphocytes exposed to OKT_4 antibody	Mechanical	11
Plastic dishes	Goat anti-mouse Ig antibodies	Mouse Ig$^+$ spleen cells	Vigorous pipetting	12
Plastic dishes	− Rabbit anti-mouse Ig antibodies − F(ab′)$_2$ fragment of rabbit anti-mouse Ig antibodies	Mouse Ig$^+$ spleen cells	Vigorous pipetting	13
Poly-acryl-amide beads	Rabbit anti-mouse IgG antibodies (covalently coupled to the beads)	Human T cell subsets labelled with mouse monoclonal OKT_4 and OKT_8 antibodies	Excess goat anti-mouse IgG	16
Sephadex beads	Anti-human F(ab′)$_2$ antibodies	Purification of human B lymphocytes (bearing surface Ig)	Excess human Ig	17,18

Agarose-poly-acrolein beads	1. Goat anti-rabbit IgG antibodies. 2. Goat anti-mouse Ig, anti-Thy-1,2, soybean agglutinin (coupled with polylysine-glutaraldehyde spacer)	1. Human erythrocytes sensitised with rabbit anti-human erythrocyte antibodies 2. Separation of B and T mouse splenocytes	Gentle stirring of the beads	19
Plastic dishes	Anti-fluorescein antibodies	Fractionation of human lymphocytes labelled with specific fluoresceinated antibodies: − helper T cells labelled with anti-Leu3a − suppressor T cells labelled with anti-Leu2a monoclonal antibodies − T-ALL cell lines labelled with anti-T-ALL antibodies − Human B lymphocytes labelled with fluoresceinated anti-human Ig	Reversal of binding with fluorescein-L-lysine + gentle aspiration	14,15
Sepharose 6 MB	Covalently linked protein A	− Mouse thymocytes charged with rabbit anti-θ Ig − Mouse spleen lymphocytes treated with rabbit anti-mouse Ig	Excess dog IgG or mechanical treatment	20
Plastic tubes	Human Ig (immobilised on plastic tubes through carbodiimide coupling) + Protein A + antibody against specific antisera	Human peripheral blood lymphocytes incubated with specific antisera	Scraping of the immunoadsorbent surface	21
Plastic flasks	IgG (coupled to the dishes by the cabodiimide procedure) + Protein A + antisera specific for cell membrane antigens (rabbit anti-allotype sera)	Rabbit lymphocyte subpopulations	Scraping of the immunoadsorbent surface	22
Plastic dishes	Haptenated gelatin (DNP-gelatin) adsorbed on the plastic	Antigen-binding spleen cells	Melting of the gelatin, then DNP-gelatin digestion by collagenase	26
Nylon discs Latex beads	DNP-immobilised in a chemically well defined manner, through S-S bonds	Antigen-binding spleen cells	− thermal − cleavage of the S-S bond with 10 mM mercaptoethanol	29
Plastic dishes covered with gelatin	Phosphorylcholine bound to gelatin plates through S-S bonds	Antigen-binding spleen cells	Cleavage of the S-S bond with 50 mM dithiothreitol	30

(iii) *Methods involving protein A from Staphylococcus aureus* (8,20 – 23). Protein A possesses the property of binding the Fc portion of IgG (Protein A is bivalent) in a theoretically reversible fashion; cell separation can be achieved with Protein A-coated immunoglobulin-specific supports: Protein A is bound either directly to beads or to immunoglobulin-coated plastic surfaces; this immobilised Protein A can either directly interact with specific antibody-coated cells or be charged with antibodies directed against the specific antibodies used to select the desired cell population.

It is generally agreed (1 – 4, references of *Table 1*) that two main types of problems are encountered in cell affinity chromatography.

(i) Non-specific interaction between cells and affinity supports.

(ii) Irreversibility of cell-affinity support interactions: although adsorption of cells to specific immobilised ligands is easily often achieved, difficulties have systematically been encountered in the recovery of bound cells by processes consistent with cell viability. One reason for this is the multipoint attachment which occurs between cell and support: each cell possesses numerous receptors for the immobilised ligand, several ligand molecules being themselves bound to the same matrix particle or surface. A second reason for irreversible binding of cells to the affinity support may be the very high affinity between the immobilised ligand and its cell surface receptor, which often occurs when cell-ligand recognition involves antigen-antibody interactions; this is a severe limitation on the isolation of viable cells by immunoaffinity chromatography.

It has sometimes been possible to recover cells initially bound to immobilised ligands by the use of excess ligand or ligand competitor (16 – 18,24), but this ideal situation is feasible only occasionally. In most cases alternatives must be found (1 – 4): melting the support (25,26), enzymatic degradation of the support (27), mechanical disruption. All these methods may damage the cells. Even pipetting by itself is liable to create shearing forces, which may cause loss of receptors from the cell surface. Jakobovitz *et al.* (28) recently took advantage of this irreversible binding of cells to affinity matrices to detach lectin receptors from erythrocytes by mechanical disruption of the immobilised lectin-bound cell complexes.

The use of ox red blood cells as a support (*Table 1*) is an interesting approach to overcome the irreversibility of cell-affinity matrix binding, since rosetting cells can be isolated by red blood cell-selective lysis.

Another attempt was that developed by Kiefer (29) and Cambier and Neale (30) who isolated antigen-specific lymphocytes by inserting a cleavable disulphide bridge between solid support and attached haptens or antigens. However, cleavage of these S-S bonds with thiol was sometimes difficult and necessitated additional mechanical assistance (30).

It is clear that to date there is no single method of cell isolation by affinity chromatography that can be applied universally, mainly because the elution of intact functional cells from the affinity supports remains a troublesome step.

3. CELL AFFINITY CHROMATOGRAPHY WITH LIGANDS IMMOBIL- ISED THROUGH CLEAVABLE MERCURY-SULPHUR BONDS

In order to try to overcome some of the disadvantages of the above methods, im- mobilisation of ligands has been achieved through spacer arms containing Hg-S bonds which can be cleaved with a thiol more readily than disulphide bridges: The organomercurial Mersalyl (sodium *o*-[3-[hydroxymercuri]-2-methoxypropyl)- carbamoyl]-phenoxyacetate) was covalently bound to activated trisacryl beads and this immobilised mercurial compound was reacted with thiolated ligands: the resulting Hg-S bonds can easily be cleaved with dithiothreitol (31).

3.1 Principle

3.1.1 *The Support*

This consists of trisacryl beads (Trisacryl GF 05, Reactifs IBF, France). The main characteristics of this macromolecule are that it bears three hydroxymethyl groups and one alkylamide group for each principal repeating unit. Because of these chemical groups, the polymer is very hydrophilic and suitable for the separation of biological macromolecules, especially proteins, and cells. This matrix has an obvious advantage over polyacrylamide- or hydroxymethylmeth- acrylate-based supports, which have a pronounced hydrophobic character. In addition, trisacryl beads do not give non-specific interactions with cells, mainly because, unlike classical chromatography matrices, they possess no saccharide structures.

3.1.2 *Activation of Trisacryl*

Primary amino groups are introduced into trisacryl beads by the action of epichlorohydrin in the presence of zinc tetrafluoroborate (32), followed by am- monia treatment, according to *Figure 1*.

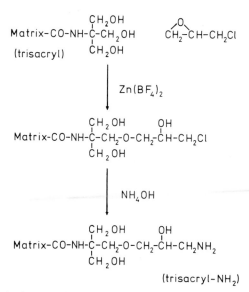

Figure 1. Introduction of primary amino groups into trisacryl beads.

195

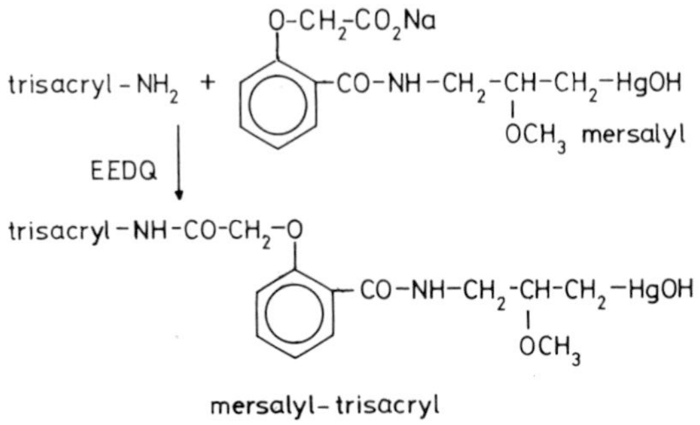

$O-CH_2-CO_2Na$

trisacryl – NH$_2$ + —CO–NH–CH$_2$–CH–CH$_2$–HgOH

OCH_3 mersalyl

EEDQ

trisacryl –NH–CO–CH$_2$–O

—CO–NH–CH$_2$–CH–CH$_2$–HgOH

OCH_3

mersalyl– trisacryl

Figure 2. Coupling of Mersalyl to trisacryl-NH$_2$.

The concentration of NH$_2$ groups, as determined by frontal analysis, is approximately 50 μmol/ml gel.

3.1.3 *Coupling of Mersalyl to Trisacryl-NH$_2$*

An amido bond between the carboxyl group of Mersalyl and the amino group of the matrix is formed by the action of N-ethoxycarbonyl-2-ethoxy-1, 2-dihydroquinoline (EEDQ) as coupling agent (33,34) (*Figure 2*). The coupling reaction is carried out in ethanol-water (1:1) mixture; the solubility of Mersalyl is convenient for this reaction, which is not the case for other classical organomercurials (such as *p*-chloromercuribenzoic acid).

The amount of bound Mersalyl, determined by the addition of [^{14}C]Mersalyl to the assays, is 0.2 μmol/ml gel. The number of accessible bound Mersalyl molecules is estimated by saturating Mersalyl-trisacryl columns with thiosalicylic acid, a thiol which possesses a chromophoric group, and then recovering the bound thiosalicylic acid by washing the column with excess dithiothreitol: this is approximately 0.1 μmol Mersalyl/ml gel, which is not very different from the value obtained with [^{14}C]Mersalyl binding. Although the amount of immobilised Mersalyl appears rather low, it is sufficient to saturate the surface of the beads with thiolated ligands having the size of lectins or immunoglobulins.

3.1.4 *Ligand Thiolation and Immobilisation of Thiolated Ligands on Mersalyl-trisacryl*

Ligands [concanavalin A (con A), antibodies] may be thiolated as described by Klotz and Heiney (35) using S-acetylmercaptosuccinic anhydride (SAMSA) as thiolating agent (*Figure 3*).

Immobilisation on Mersalyl-trisacryl through an Hg-S bond may be carried out as indicated in *Figure 3*. Initial immobilisation experiments (31) were carried out by simultaneously mixing the S-acetyl ligand derivative, Mersalyl-trisacryl and hydroxylamine to generate *in situ* the -SH functions of the modified ligand (35). However, the S-acetylated ligand can react directly with immobilised Mersalyl without hydroxylamine reduction of the S-acetyl group (31), thus avoiding possible oxidation of -SH groups.

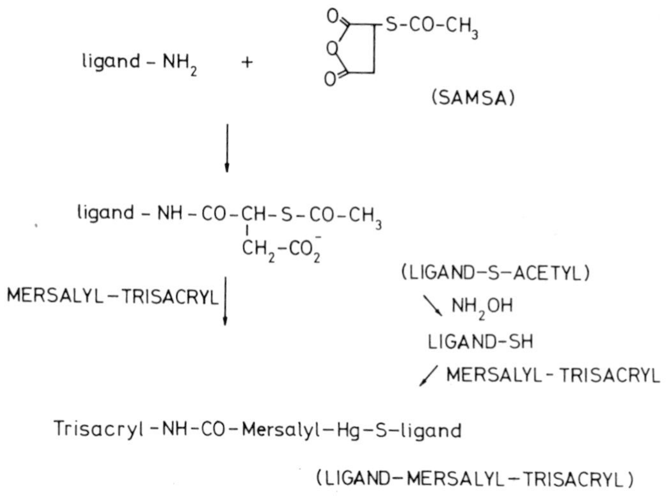

Figure 3. Immobilisation of ligands on Mersalyl-trisacryl.

The degree of ligand thiolation [monitored by hydroxylamine treatment of the S-acetyl derivative (35), followed by classical -SH determination (36)] can be controlled by judicious adjustment of ligand and SAMSA concentrations. When the degree of ligand thiolation is too high, the recovery of cells initially bound to the resulting support is much more difficult: this is probably due to additional multipoint attachment between the ligand and the Mersalyl molecules of the support. The extent of immobilised ligand can be determined by the use of radioactive ligands ([³H]concanavalin A, ¹²⁵I-labelled IgG).

3.1.5 *Cell Binding to Ligand-Mersalyl-trisacryl Supports. Recovery by Thiol Treatment*

Cells possessing surface components recognising the immobilised ligand are incubated with ligand Mersalyl-trisacryl gel, and various control gels. Recovery of bound cells is achieved by thiol treatment. The principle of cell isolation using ligand-Mersalyl-trisacryl is summarised in *Figure 4*.

Cells are recovered with ligand molecules which are now monovalently bound to the cell surface. These ligand molecules can either be eliminated with competitive agents (thus, in the case of lectins, complexes can be broken by sugars) or taken up into the cells by endocytosis (in the case of antibodies).

3.2 **Methods**

3.2.1 *Introduction of Amino Groups on Trisacryl Beads*

(i) Treat trisacryl beads (Trisacryl GF 05, Reactifs IBF) (30 ml), with 25% (w/v) zinc tetrafluoroborate (22.5 ml) and epichlorohydrin (60 ml), for 3 h at 80°C agitating gently.

(ii) Cool the reaction mixture, rinse the beads with distilled water and resuspend them in 2.0 M ammonia-ammonium chloride, pH 9.0 (100 ml).

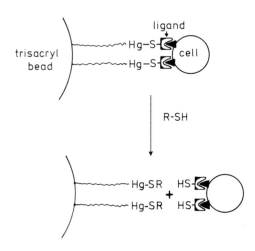

Figure 4. The principle of cell isolation using ligand-Mersalyl-trisacryl.

(iii) After 16 h gentle agitation at room temperature, rinse the beads thoroughly with distilled water.

3.2.2 Immobilisation of Mersalyl on Trisacryl-NH₂ Beads

The coupling is carried out reaction in an ethanol-water (1:1) mixture according to the method developed by Boschetti *et al.* (33).

(i) Add water (13 ml) and ethanol (27.5 ml) to sedimented beads (17 ml).

(ii) Dissolve Mersalyl (Sigma) (200 mg) in 0.15 M NaOH (3.3 ml), and adjust the pH to 7.5 with 2.0 M HCl and add ethanol (3.4 ml).

(iii) Quickly pour this Mersalyl solution into ethanol (2.5 ml) containing EEDQ (Aldrich) (300 mg).

(iv) Shake vigorously for 2 min and add the EEDQ-Mersalyl mixture to the trisacryl-NH₂ beads.

(v) Rotate the mixture end-over-end overnight at room temperature.

(vi) Thoroughly rinse the beads with ethanol-water (1:1) (1.5 litres), water (1 litre) and 1.0 M NaCl (1 litre).

Mersalyl-trisacryl should be protected from light; it may be stored frozen at −20°C.

[^{14}C]Carbamoyl-Mersalyl (CEN, Saclay) can be included in the experiment to allow evaluation of the amount of immobilised Mersalyl. An alternative method is as follows: equilibrate the gel column in 50 mM phosphate, pH 6.2, containing 0.15 M NaCl and 1% (v/v) Lubrol PX, and saturate with 20 mM thiosalicylic acid; elute the bound thiosalicylic acid with 20 mM dithiothreitol in the buffer described above and determine the amount of thiosalicylic acid eluted by ultraviolet spectrophotometry.

3.2.3 Ligand Thiolation

(i) To 1 ml of a ligand solution containing 1 – 2 mg lectin (or antibody) per ml (in 125 mM phosphate buffer, pH 7.0) add freshly prepared SAMSA

(Sigma) solution (50 – 100 mM) in dimethylformamide (20 μl SAMSA solution per ml of ligand solution).
(ii)   Gently rotate the mixture for 2 h at room temperature.
(iii) Eliminate unreacted SAMSA either by dialysis against 125 mM phosphate buffer, pH 7.0, or by gel filtration through a small column of trisacryl GF 05 equilibrated in the same buffer.
(iv)   To measure the degree of thiolation of the ligand, treat the S-acetyl ligand derivative with hydroxylamine (at a final concentration of 20 mM) for 1 h at room temperature and determine the -SH content by the method of Grassetti and Murray (36): add an equal volume of 2 mM dithiodipyridine in 50 mM phosphate buffer, pH 7.0, and read the increase in optical density at 343 nm (the molar extinction coefficient of thiopyridone, the reaction product, is 8.08×10^3). This measurement must be carried out immediately after addition of the dithiodipyridine solution because of the slow reaction of the latter with hydroxylamine.

3.2.4 *Ligand Immobilisation on Mersalyl-trisacryl*

(i)   Mix S-acetyl ligand derivative (0.5 – 2 mg/ml in 125 mM phosphate buffer, pH 7.0) with an equal volume of packed Mersalyl-trisacryl beads in the same buffer.
(ii)   Rotate end-over-end for 12 – 16 h at room temperature in the dark.
(iii) Thoroughly rinse the gel with phosphate buffer, then with 10 mM Tris-HCl pH 7.5, containing 0.5 M NaCl.
(iv)   Resuspend the beads in the medium chosen for cell binding.

Control experiments show that non-thiolated ligand cannot be fixed on the beads. However, with some batches of antibody a very weak fixation (visualised by subsequent cell binding) may occur through non-specific interactions or undesired (unknown) reactions between Mersalyl and proteins; this can be suppressed by including bovine serum albumin (0.5% w/v) in the immobilisation procedures.

3.2.5 *Binding of Cells to Ligand-Mersalyl-trisacryl Gels: Recovery of Bound Cells by Thiol Treatment*

As columns appear to trap significant numbers of cells mechanically, a batchwise procedure is preferable.
(i)   Disperse the ligand-Mersalyl-trisacryl gel (or control gel) in a large surface vessel allowing spreading of the beads.
(ii)   Incubate the gels with the cell suspension, with occasional very gentle agitation.
(iii) After elimination of unbound cells and repeated gentle washing, transfer some beads to a counting chamber for microscope examination.
(iv)   To recover bound cells treat the beads with thiol supplemented medium, with occasional gentle agitation.

3.3 **Applications**

Three model studies demonstrating the feasibility of this methodology are described.

3.3.1 *Thymocyte Binding to Con A-Mersalyl-trisacryl. Elution with Dithiothreitol*

Initial experiments were carried out with SPDP-thiolated Con A (31), although it has since been found that the SAMSA procedure was more convenient. Con A (Pharmacia) solutions (2 mg/ml) are thiolated as indicated in Section 3.2 using 1 – 2 mM SAMSA (final concentration); this results in the introduction of 1.5 – 3.3 S-acetyl groups per Con A molecule. The amount of immobilised Con A, measured with [³H]Con A (NEN), is approximately 60 μg per ml gel.

Mouse thymocytes are incubated with Con A-Mersalyl-trisacryl, or various control gels. After 20 min incubation, followed by washing to remove unbound cells, the beads are examined under a light microscope. The results are shown in *Figure 5*.

Figure 5a shows results with the control gels. In all cases there is no significant binding of thymocytes to the beads, showing that the trisacryl-derived matrix has no significant non-specific interaction with the cells and that neither Con A-S-Ac nor non-thiolated Con A are non-specifically adsorbed to the beads.

Figure 5b shows Con A-Mersalyl-trisacryl beads, prepared as described above, after incubation with mouse thymocytes: the beads are covered with cells. Bound cells cannot be eluted by 2% (w/v) methyl-α-D-mannopyranoside, which specifically binds to Con A.

Recovery of bound thymocytes is readily achieved by treatment with 50 mM dithiothreitol containing methyl-α-D-mannopyranoside (2% w/v, added to prevent Con A-induced agglutination of released cells). *Figure 5c* shows thymocytes dissociated from such a support after 30 min treatment with this elution mixture. About 70% of the bound cells are eluted within 30 min and all the cells may be recovered after longer times. Dithiothreitol eluted thymocytes may be shown to be viable since they exclude trypan blue and are stimulated normally by Con A, as demonstrated by [³H]thymidine incorporation experiments (31).

3.3.2 *Binding of TNBS-Labelled Erythrocytes to Anti-DNP-Mersalyl-trisacryl. Recovery by Thiol Treatment*

This study consists of the immobilisation of thiolated monoclonal anti-DNP antibodies on Mersalyl-trisacryl, and the binding of TNBS-labelled sheep erythrocytes which are recognised by anti-DNP antibodies.

(i)   Thiolate 1 mg/ml solution of monoclonal anti-DNP antibody (Clin Midy-SANOFI Research Center, Montpellier) as indicated in Section 3.2 using 1 – 2 mM SAMSA (final concentrations); this results in the introduction of 1.7 – 3.5 S-acetyl groups per antibody molecule. The amount of bound antibody, determined with ¹²⁵I-labelled antibodies, should be of the order of μg per ml gel.

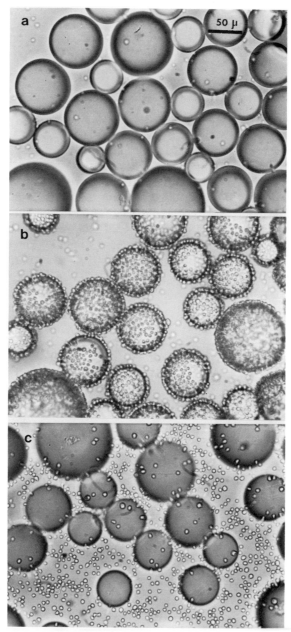

Figure 5. Binding of mouse thymocytes to Con A-Mersalyl-trisacryl. Mouse thymocytes were incubated for 20 min at room temperature with Con A-Mersalyl-trisacryl gels or various control gels (5 x 10^7 cells in 250 μl 10 mM Tris-HCl, pH 7.5, 0.15 M NaCl, per 100 μl gel). After removal of unbound cells by repeated washing, the beads were examined under a light microscope. **a.** Control gels (trisacryl, or trisacryl-NH_2, or Mersalyl-trisacryl without Con A, or Mersalyl-trisacryl pre-treated with non-thiolated Con A, or trisacryl-NH_2 pre-treated with Con A-S-Ac). **b.** Con A-Mersalyl-trisacryl. **c.** Con A-Mersalyl-trisacryl after thymocyte binding and treatment by 50 mM dithiothreitol, 2% methyl-α-D-mannopyranoside, in 10 mM Tris-HCl, pH 7.5, 0.15 M NaCl, for 30 min at room temperature with very gentle agitation.

(ii)   Label sheep erythrocytes with trinitrobenzene sulphonic acid (TNBS) as described by Rittenberg and Pratt (37).

(iii) Incubate TNBS-labelled sheep erythrocytes with anti-DNP-Mersalyl-trisacryl gels or various control gels in large surface vessels with occasional gentle agitation.

(iv)   After 20 min incubation and removal of unbound cells by several washings, examine the beads under a light microscope.

Figure 6a shows anti-DNP-Mersalyl-trisacryl beads which have been incubated with unlabelled erythrocytes; these do not bind to the support. *Figure 6b* shows TNBS-labelled erythrocytes incubated with Mersalyl-trisacryl pre-treated with non-thiolated anti-DNP: no cell binding occurred, showing that no antibody was non-specifically adsorbed to the beads. *Figure 6c* shows anti-DNP-Mersalyl-trisacryl beads incubated with TNBS-labelled erythrocytes; the beads are completely covered with erythrocytes, which cannot be eluted by washing. *Figure 6d* shows dissociation of TNBS-labelled erythrocytes initially bound to anti-DNP-Mersalyl-trisacryl beads after 30 min treatment with 50 mM dithiothreitol. All the cells are eluted and no haemolysis is observed. The number of erythrocytes bound to the matrix which are eluted by thiol cleavage of the Hg-S bond is 10^9 per ml gel.

3.3.3 *Binding of Mouse Thymocytes to Anti-Thy-1,2-Mersalyl-trisacryl. Elution with Dithiothreitol*

First results have been obtained with immobilised antibodies recognising natural cell surface antigens (38).

(i)   Thiolate a 2 mg/ml solution of monoclonal anti-Thy-1,2 antibody (Clin-Midy-SANOFI Research Center, Montpellier) as indicated in Section 3.2 by 1 mM SAMSA (final concentration).

(ii)   React the resulting S-acetyl derivative for 16 h at room temperature, in the dark, with an equal volume of Mersalyl-trisacryl.

(iii) Incubate anti-Thy-1,2-Mersalyl-trisacryl gel with mouse thymocytes for 2 h at 4°C in Hank's medium containing 0.1% (w/v) BSA and 0.1% (w/v) sodium azide.

Figure 7a shows the binding of thymocytes from a Thy-1,2-positive mouse strain (C57/BL6): these thymocytes were recovered in viable state by 2 h treatment with 25 mM dithiothreitol in the medium indicated above (*Figure 7b*). The thymocyte binding to anti-Thy-1,2-Mersalyl-trisacryl was specific since thymocytes from a Thy-1,1-positive mouse strain (C57/KA from Dr Bonnier's laboratory, Liege) were not adsorbed (*Figure 7c*).

4. CONCLUDING REMARKS

The development of cell separation by affinity chromatography techniques has been severely hampered by difficulties in recovering cells from affinity supports. Although various proposals have been made to solve these problems, no single method has yet been found to be universally applicable. This chapter has outlined

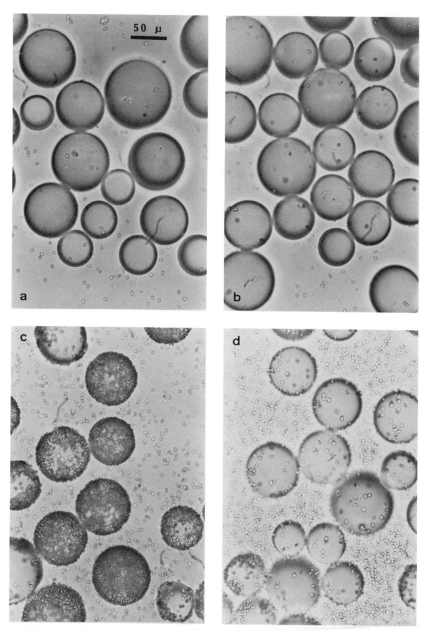

Figure 6. Binding of TNBS-labelled erythrocytes to anti-DNP-Mersalyl-trisacryl. Sheep erythrocytes are incubated for 20 min at room temperature with anti-DNP-Mersalyl-trisacryl gels or various control gels (2 x 10^8 cells in 250 μl 10 mM Tris-HCl, pH 7.5, 0.15 M NaCl, per 100 μl gel), in large surface vessels, with very gentle occasional agitation. After removal of unbound cells by washing, the beads are examined under a light microscope. **a**. Anti-DNP-Mersalyl-trisacryl beads + unlabelled erythrocytes. **b**. TNBS-labelled erythrocytes + Mersalyl-trisacryl beads pre-treated with non-thiolated anti-DNP. **c**. Anti-DNP-Mersalyl-trisacryl beads + TNBS-labelled erythrocytes. **d**. Dissociation TNBS-labelled erythrocytes, initially bound to anti-DNP-Mersalyl-trisacryl beads, after 30 min treatment with 50 mM dithiothreitol in 10 mM Tris-HCl, 0.12 M NaCl, pH 7.5.

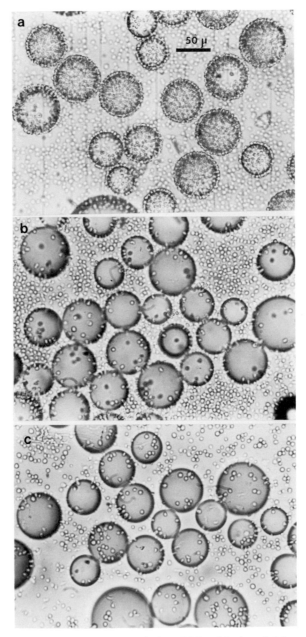

Figure 7. Binding of mouse thymocytes to anti-Thy-1,2-Mersalyl-trisacryl. Mouse thymocytes were incubated for 2 h at 4°C with anti-Thy-1,2-Mersalyl-trisacryl in Hank's medium containing 0.1% (w/v) BSA and 0.1% (w/v) sodium azide (10^8 cells in 250 μl medium per 50 μl gel). After removal of unbound cells by washing, the beads were examined under a light microscope. **a.** Anti-Thy-1,2-Mersalyl-trisacryl + thymocytes of a Thy-1,2-positive mouse strain (C57/BL6). **b.** Anti-Thy-1,2-Mersalyl-trisacryl + Thy-1,2-postive thymocytes, after 2 h treatment with 25 mM dithiothreitol in Hank's medium. **c.** Anti-Thy-1,2-Mersalyl-trisacryl + thymocytes of a Thy-1,1-positive mouse strain (C57/KA).

a method utilising cleavable spacer arms: the matrix is composed of trisacryl beads, which do not interact with cells. The organomercurial Mersalyl is covalently immobilised to the matrix by a simple chemical reaction, and ligands thiolated with a commercially available heterobifunctional reagent (SAMSA) are reacted with immobilised Mersalyl: the resulting Hg-S bond may be cleaved more readily than disulphide bridges with a thiol, which is especially important when there is a possibility of multivalent binding of the thiolated ligand to the matrix.

Preliminary results suggest that immobilisation on Mersalyl-trisacryl of anti-mouse immunoglobulins and isolation of cells previously coated with specific mouse monoclonal antibodies can be applied as a general method. This process, already used in 'panning techniques', requires only small quantities of these monoclonal antibodies (Bonnafous *et al.*, unpublished observations). This technique promises new applications in the separation of specific cell subsets and in the isolation of antigen-specific lymphocytes.

5. ACKNOWLEDGEMENTS

We thank Drs H.Vidal and P.Poncelet (Clin-Mindy-SANOFI Research Center, Montpellier) for a gift of monoclonal antibodies and helpful discussions, and R. Larguier and N. Bernad for skilful technical assistance. This work received financial support from the Direction du Développement Scientifique et Technologique et de l'Innovation (Grant no. 82.V.0016), the Centre National de la Recherche Scientifique and the Fondation pour la Recherche Médicale Française.

6. REFERENCES

1. Edelman,G.M. and Rutishauser,U. (1974) in *Methods in Enzymology*, Vol. **34**, Jakoby,N.B. and Wilchek,M. (eds.). Academic Press, New York, p. 195.
2. Haas,W. and von Boehmer,H. (1978) in *Current Topics in Microbiology and Immunology: Techniques for Separation and Selection of Antigen Specific Lymphocytes*, Haas,W. and von Boehmer,H. (eds.). Springer Verlag, New York, p. 1.
3. Sharma,S.K. and Mahendroo,P.P. (1980) *J. Chromatogr.*, **184**, 471.
4. Basch,R.S., Berman,J.W. and Lakow,E. (1983) *J. Immunol. Methods*, **56**, 269.
5. Stocker,J.W., Garotta,G., Hausmann,B., Trucco,M. and Ceppellini,R. (1979) *Tissue Antigens*, **13**, 212.
6. Egeland,T. and Lea,T. (1982) *J. Immunol. Methods*, **55**, 213.
7. Mills,K., Armitage,R. and Worman,C. (1983) *Immunol. Lett.*, **6**, 241.
8. Karavodin,L.M. and Golub,S.H. (1983) *J. Immunol. Methods*, **61**, 293.
9. Tsoi,M.S., Aprile,J., Dobbs,S., Goehle,S. and Storb,R. (1982) *J. Immunol. Methods*, **53**, 293.
10. Reinherz,E.L., Penta,A.C., Hussey,R.E. and Schlossman,S.F. (1981) *Clin. Immunol. Immunopathol.*, **21**, 257.
11. Payne,S.M., Sharrow,S.O., Shearer,G.M. and Biddison,W.E. (1981) *Int. J. Immunopharmacol.*, **3**, 227.
12. Mage,M.G., McHugh,L.L. and Rothstein,T.L. (1977) *J. Immunol. Methods*, **15**, 47.
13. Wysocki,L.J. and Sato,V.L. (1982) *Proc. Natl. Acad. Sci. USA*, **75**, 2844.
14. Fong,S., Tsoukas,C.D., Pasquali,J.L., Fox,R.I., Rose,J.E., Raiklen,D., Carson,D.A. and Vaughan,J.H. (1981) *J. Immunol. Methods*, **44**, 171.
15. Fong,S., Fox,R.I., Rose,J.E., Liu,J., Tsoukas,C.D., Carson,D.A. and Vaughan,J.H. (1981) *J. Immunol. Methods*, **46**, 153.
16. Braun,R., Teute,H., Kirchner,H. and Munk,K. (1982) *J. Immunol. Methods*, **54**, 251.
17. Chess,L., McDermott,R.P. and Schlossman,S.F. (1974) *J. Immunol.*, **113**, 1113.

18. Anderson,K.C., Griffin,J.D., Bates,M.P.M., Slaughenhoupt,B.L., Schlossman,S.F. and Nadler,L.M. (1983) *J. Immunol. Methods,* **61**, 283.
19. Margel,S., Ofarim,M. and Eshhar,Z. (1983) *J. Cell Sci.,* **62**, 149.
20. Gheti,V., Mota,G. and Sjoquist,J. (1978) *J. Immunol. Methods,* **21**, 133.
21. Bundesen,P.G. and Gordon,J. (1979) *J. Immunol. Methods,* **30**, 179.
22. Nash,A.A. (1979) *J. Immunol. Methods,* **12**, 149.
23. Langone,J.J. (1982) *J. Immunol. Methods,* **55**, 277.
24. Hellstrom,U., Hammarstrom,S., Dillner,M.L., Perlmann,H. and Perlmann,P. (1976) *Scand. J. Immunol.,* **5**, 45.
25. Rutishauser,U., d'Eustachio,P. and Edelman,G.M. (1973) *Proc. Natl. Acad. Sci. USA,* **70**, 3894.
26. Haas,W. and Layton,J.E. (1975) *J. Exp. Med.,* **141**, 1004.
27. Schlossman,S.F. and Hudson,L. (1973) *J. Immunol.,* **110**, 149.
28. Jakobovitz,A., Eshdat,Y. and Sharon,N. (1981) *Biochem. Biophys. Res. Commun.,* **100**, 1484.
29. Kiefer,H. (1975) *Eur. J. Immunol.,* **5**, 624.
30. Cambier,J.C. and Neale,M.J. (1982) *J. Immunol. Methods,* **51**, 209.
31. Bonnafous,J.C., Dornand,J., Favero,J., Sizes,M., Boschetti,E. and Mani,J.C. (1983) *J. Immunol. Methods,* **58**, 93.
32. Hubert,P., Mester,J., Dellacherie,E., Neel,J. and Beaulieu,E.E. (1978) *Proc. Natl. Acad. Sci. USA,* **75**, 3143.
33. Boschetti,E., Corgier,M. and Garelle,R. (1978) *Biochimie,* **60**, 425.
34. Pougeois,R., Satre,M. and Vignais,P.V. (1978) *Biochemistry (Wash.),* **17**, 3018.
35. Klotz,I.M. and Heiney,R.E. (1962) *Arch. Biochem. Biophys.,* **96**, 605.
36. Grassetti,D.R. and Murray,J.F.,Jr. (1967) *Arch. Biochem. Biophys.,* **119**, 41.
37. Rittenberg,M.B. and Pratt,K.L. (1969) *Proc. Soc. Exp. Biol. Med.,* **132**, 575.
38. Bonnafous,J.C., Dornand,J., Favero,J., Mani,J.C. and Boschetti,E. (1983) in *Affinity Chromatography and Biological Recognition,* Chaiken,I.M., Wilchek,M. and Parikh,I. (eds.) Academic Press Inc., London and New York, in press.

INDEX

Forthcoming

Animal cell culture
a practical approach

Edited by R I Freshney

After an introductory chapter dealing with basic techniques, this book provides detailed protocols both for traditional areas like organ culture, characterisation and storage, and for developing areas such as serum-free media, cell separation and *in situ* hybridisation.

January 1986; 250pp (approx); softbound: *0 947946 33 0;*
£14.00/US$25.00; hardbound: *0 947946 62 4; £22.00/US$40.00*

Photosynthetic energy transduction
a practical approach

Edited by M F Hipkins and N R Baker

An up-to-date laboratory manual for researchers and students wishing to learn a wide range of techniques for the study of photosynthetic energy transduction.

Due 1986; 250pp (approx); softbound: *0 947946 51 9;* hardbound:
0 947946 63 2

Biochemical toxicology
a practical approach

Edited by K Snell and B Mullock

Chapters written by laboratory experts provide practical guidance and 'tricks of the trade' for the most useful techniques in toxicological research. The book is unique as a guide for researchers at all levels, especially those in pharmaceutical and agrochemical laboratories.

Due 1986; 250pp (approx); softbound: *0 947946 52 7;* hardbound:
0 947946 65 5

Spectrophotometry and spectrofluorimetry
a practical approach
Edited by D A Harris and C L Bashford

Due 1986; 250pp (approx); softbound: *0 947946 46 2;* hardbound:
0 947946 69 1

PRICES TO BE ANNOUNCED

⬡ **IRL PRESS**   IRL Press Ltd, PO Box 1, Eynsham, Oxford OX8 1JJ, UK
IRL Press Inc, PO Box Q, McLean, VA 22101, USA